Endogenous Development

Western ideas, worldviews, actors, tools, models, and frameworks have long dominated development theory and practice in Africa. The resulting development interventions are too rarely locally rooted, locally driven, or resonant with local context. At the same time, theories and practices from developing countries rarely travel to the Western agencies dominating development, undermining the possibility of a beneficial synergy that could be obtained from the best of both worlds. There are many reasons why the experiences of locally driven development are not communicated back to global development actors, including, but not limited to, the marginal role of Southern voices in global forums.

This volume gives a platform to authentic African voices and non-African collaborators, to explore what endogenous development means, how it can be implemented, and how an endogenous development approach can shape local, national and global policies.

This book was originally published as a special issue of *Development in Practice*.

Chiku Malunga is an Organizational Paremiologist. He works with civil society organizations as an Organization Development Practitioner, using African indigenous wisdom in African proverbs and folktales to transform development management.

Susan H. Holcombe has worked with Oxfam America and several United Nations organizations, and has taught Sustainable Development at Brandeis University, Waltham, Massachusetts, USA. Her career ranges from on the ground experience in Africa and Asia, to teaching and research.

Development in Practice Books
Series Editor: Brian Pratt

Each title in the *Development in Practice Books* series offers a focused overview of practice-relevant analysis, experience, and research on key topics in development.

Endogenous Development

Naïve romanticism or practical route to
sustainable African development

Edited by
Chiku Malunga and Susan H. Holcombe

Routledge
Taylor & Francis Group
LONDON AND NEW YORK

International NGO Training and Research Centre

First published 2016
by Routledge
2 Park Square, Milton Park, Abingdon, Oxon, OX14 4RN, UK

and by Routledge
711 Third Avenue, New York, NY 10017, USA

First issued in paperback 2017

Routledge is an imprint of the Taylor & Francis Group, an informa business

British Library Cataloguing in Publication Data
A catalogue record for this book is available from the British Library

ISBN 13: 978-1-138-29495-0 (pbk)
ISBN 13: 978-1-138-93680-5 (hbk)

Typeset in Times New Roman
by RefineCatch Limited, Bungay, Suffolk

Publisher's Note
The publisher accepts responsibility for any inconsistencies that may have
arisen during the conversion of this book from journal articles to book chapters,
namely the possible inclusion of journal terminology.

Disclaimer
Every effort has been made to contact copyright holders for their permission to
reprint material in this book. The publishers would be grateful to hear from any
copyright holder who is not here acknowledged and will undertake to rectify
any errors or omissions in future editions of this book.

Contents

CONTENTS

Citation Information

The chapters in this book were originally published in *Development in Practice*, volume 24, issue 5–6 (August 2014). When citing this material, please use the original page numbering for each article, as follows:

Chapter 1: Guest Editors' Introduction
Endogenous development: naïve romanticism or practical route to sustainable African development?
Chiku Malunga and Susan H. Holcombe
Development in Practice, volume 24, issue 5–6 (August 2014) pp. 615–622

Chapter 2
Identifying and understanding African norms and values that support endogenous development in Africa
Chiku Malunga
Development in Practice, volume 24, issue 5–6 (August 2014) pp. 623–636

Chapter 3
Endogenous development: some issues of concern
David Millar
Development in Practice, volume 24, issue 5–6 (August 2014) pp. 637–647

Chapter 4
African family values in a globalised world: the speed and intensity of change in post-colonial Africa
Charles Banda
Development in Practice, volume 24, issue 5–6 (August 2014) pp. 648–655

Chapter 5
African philanthropy, pan-Africanism, and Africa's development
Bhekinkosi Moyo and Katiana Ramsamy
Development in Practice, volume 24, issue 5–6 (August 2014) pp. 656–671

Chapter 6
Wiki approaches to wicked problems: considering African traditions in innovative collaborative approaches
Dawn S. Booker
Development in Practice, volume 24, issue 5–6 (August 2014) pp. 672–685

Chapter 16

Endogenous development going forward: learning and action
Chiku Malunga and Susan H. Holcombe
Development in Practice, volume 24, issue 5–6 (August 2014) pp. 777–781

For any permission-related enquiries please visit:
http://www.tandfonline.com/page/help/permissions

Notes on Contributors

Kolawole Adebayo is a Reader at the Federal University of Agriculture, Abeokuta, Nigeria, and manages the Cassava: Adding Value for Africa (C:AVA) on secondment to the Natural Resources Institute, University of Greenwich, Chatham, UK. His interests include the uptake and dissemination of agricultural innovations in smallholder farming systems, and management and sustainable funding of agricultural development, as well as rural livelihoods and management of the environment. He won the Commonwealth Split-site Doctoral Scholarship (used at the University of Reading, UK in 2001); two consecutive International Foundation for Science (IFS) Research Grants between 2004 and 2006; as well as the World Bank Development Marketplace Competitive Grant in 2008.

Charles Banda is a Malawian organisational development practitioner working with CADECO (Capacity Development Consultants). He has worked with Action Aid Malawi and Concern Universal–Malawi. He has many years of experience in facilitating financial and organisational sustainability interventions among non-governmental organisations. He is the co-author (with Chiku Malunga) of *Understanding Organizational Sustainability through African Proverbs* (2011).

Dawn S. Booker is affiliated with the American University of Paris, France.

Francis Issahaku Malongza Bukari is a Lecturer in the Department of Environment and Resource Studies, Faculty of Integrated Development Studies, University for Development Studies, Tamele, Ghana.

Ariel Delaney is currently a Project Officer for African Development Solutions's Natural Resources Management (NRM) project based in Garowe, Somalia. NRM is a joint endeavour with CARE International as well as the Puntland Ministry of Environment, Wildlife, and Tourism. NRM aims to improve rangeland conditions and promote the sustainable use of resources by empowering communities and enabling the institutional and legal framework for rangeland protection.

Danielle Fuller-Wimbush is a PhD Candidate in Global Health and Development Policy at Brandeis University, Waltham, MA, USA. She is a short-term consultant for the World Bank and conducted the initial assessment of the cassava drying initiative to discern its potential for scaling. Her research focuses on sustainable agriculture, food security, aid effectiveness, and health system strengthening.

Sylvester Zackaria Galaa is Senior Lecturer and Dean of Faculty in the Department of Social, Political, and Historical Studies, Faculty of Integrated Development Studies, University for Development Studies, Tamale, Ghana.

Susan H. Holcombe is a Professor in the Heller School for Social Policy and Management at Brandeis University, Waltham, MA, USA. Her teaching and publications build on a career of practice and a focus on building capabilities for human development. She was Programme Director for Oxfam America and has served in various positions with UNFPA, UNIFEM, and UNICEF in New York and in multiple field postings. She has participated in or led field evaluations and assessments for Ford Foundation, UNFPA, UNICEF, UNDP, World Bank, and the University of the South Pacific. She currently teaches in the Brandeis MA Programme in Sustainable International Development. She also assists the The Poverty Alleviation Fund with programme planning and monitoring.

Marion Keim-Lees is currently the Director of the Interdisciplinary Centre of Excellence for Sports Science and Development (ICESSD) at the University of Western Cape, Cape Town, South Africa.

Mariama Khan is an Africanist scholar, currently completing her PhD at the Centre of African Studies, University of Edinburgh, UK. She has worked in development and the creative arts in Africa.

Chiku Malunga is the Director of CADECO (Capacity Development Consultants). He is a thought leader, author, and consultant in indigenous wisdom-based organisation development (Organisational Paremiology). His goal is to promote the African indigenous wisdom (in African proverbs, folktales, and indigenous concepts) as a tool for enhancing modern life. He has authored works on personal and organisational development using African indigenous wisdom and proverbs, and on leadership and NGO management, as well as on development in Africa more generally.

David Millar is a Fellow of the Africa Study Centre in Leiden, The Netherlands. He is also the Commonwealth Focal Person on Indigenous Peoples and Rural Communities for Ghana. He has just started a Private University in the north of Ghana with a focus on culture and development – culture more as a science of a people – to give meaning to his over 30 years' work and numerous publications in this field.

Bhekinkosi Moyo is Executive Director at the Southern Africa Trust. He was previously Programme Director for TrustAfrica. He holds a PhD in Political Science from the University of the Witwatersrand, Johannesburg, South Africa. He serves on the boards of Worldwide Initiatives for Grantmaker Support (WINGS) and the African Grantmakers Network, and has worked in the fields of governance, development, and philanthropy in Africa.

Chimwemwe A.P.S. Msukwa is a freelance consultant and development practitioner based in Malawi, who has worked in East and Southern Africa. His work has mostly focused on facilitating transformative processes in communities and organisations. He is currently working on concepts for mainstreaming violent conflict prevention and transformation in development work.

Katiana Ramsamy is a Project Coordinator at the Southern Africa Trust, which she joined to help the Trust implement its human security and youth violence initiatives. She holds a Masters and Honours degree in International Relations, and a Bachelor of Science degree in Politics and Philosophy from the University of Cape Town, South Africa, where she has also worked as a tutor and assistant lecturer.

Angélique K. Rwiyereka is a medical doctor who has worked as a clinician and as Director General in the Rwandan Ministry of Health in charge of clinical services. She initiated health programmes within the first millennium village project in Rwanda. As such, she has been involved with both policy-making and policy implementation. Most recently, she completed a doctoral programme at the Heller School for Social Policy and Management, Brandeis University, Waltham, MA, USA, where she used cost effectiveness and other techniques to compare the impact of different approaches to financing health care delivery.

Nathalie Tinguery is the Country Program Coordinator at the US African Development Foundation, Burkina Faso Field Office. She has substantial experience working with communities on livelihoods and sustainability.

Afia S. Zakiya holds a PhD in Political Science from Clark Atlanta University, Atlanta, Georgia, USA, and has 20 years' of experience in African international affairs and development practice, education, and leadership. Her scholarly writings focus on African history, politics, culture and indigenous knowledge, pan-Africanism, gender studies, WASH, ecology and climate change, and globalisation. She is an expert change-management and organisational development practitioner.

INTRODUCTION

Endogenous development: naïve romanticism or practical route to sustainable African development?

Chiku Malunga and Susan H. Holcombe

Development theory and practice in developing countries are dominated by the power of Western ideas, worldviews, actors, tools, models, and frameworks. Consequently, the resulting development interventions may too rarely be locally rooted, locally driven, or resonant with local context. Another reality is that theories and practice from developing countries rarely travel to the Western agencies dominating development, undermining the possibility of a beneficial synergy that could be obtained from the best of the two worlds: West and developing countries. There are many reasons why the experience of locally driven development is not communicated back to global development actors, including but not limited to the marginal role of Southern voices in global fora. Perhaps the greatest unwelcome and unintended outcome is that by trying to create, or perhaps better said, "clone" development in developing countries in the image of Western "development", development efforts defeat their own purpose through undermining their own relevance, legitimacy, and sustainability.

La théorie et les pratiques du développement dans les pays en voie de développement sont dominées par le pouvoir des idées, de la conception du monde, des acteurs, des outils, des modèles et des cadres occidentaux. Ainsi, les interventions de développement qui en découlent peuvent être trop rarement ancrées au niveau local et pas assez souvent impulsées par le niveau local ou en mesure de trouver un écho dans le contexte local. Une autre réalité est le fait que les théories et les pratiques émanant des pays en développement parviennent rarement aux agences occidentales qui dominent le développement, ce qui nuit à la possibilité d'une synergie bénéfique qui pourrait être tirée des meilleurs aspects des deux mondes : les pays de l'Ouest et les pays en développement. Il y a de nombreuses raisons pour lesquelles l'expérience du développement impulsé par le contexte local n'est pas communiquée aux acteurs de développement mondiaux, notamment (mais sans s'y limiter) le rôle marginal joué par les voix du Sud dans les forums mondiaux. Le résultat qui est peut-être le moins bienvenu et désiré est que, en tentant de créer, ou peut-être plutôt de « cloner » le développement dans les pays en développement à l'image du « développement » occidental, les efforts de développement vont à l'encontre de leur propre objectif en minant leurs propres pertinence, légitimité et durabilité.

En los países en desarrollo, la teoría y la práctica vinculadas al desarrollo se encuentran subsumidas a las ideas, la cosmovisión, los actores, las herramientas, los modelos y los marcos de referencia de Occidente. Por consiguiente, muy pocas veces las acciones de desarrollo resultantes se encuentran arraigadas localmente, o son impulsadas desde la base o vistas como relevantes para el contexto local. Asimismo, las teorías y las prácticas que se originan en estos países casi nunca trascienden a las agencias occidentales que dominan el ámbito del desarrollo, hecho que socava cualquier posibilidad de crear una sinergia positiva que posibilite la mejor incorporación tanto de Occidente como de los países en desarrollo.

Muchos argumentos explican las razones por las cuales las vivencias surgidas del desarrollo impulsado desde la base no inciden en los actores del desarrollo a nivel mundial, entre ellas, el rol marginal desempeñado por las voces del Sur en los foros mundiales. Por lo que, la consecuencia más desagradable y posiblemente no intencionada del esfuerzo destinado a impulsar o, mejor dicho, «clonar» el desarrollo occidental en los países en desarrollo, sea que tales acciones resultan contraproducentes, pues debilitan su propia relevancia, legitimidad y sostenibilidad.

What has gone fundamentally wrong with present-day development approaches is that they have replaced rather than built on the endogenous ways of dealing with development issues. Being locally rooted and driven means that development efforts are grounded in and inspired by local context, worldviews, values, and priorities, enabling development efforts to be relevant, legitimate, and sustainable. This article introduces this special issue of *Development in Practice*, and it also reflects the ongoing dialogue between the co-editors about the meanings and importance of endogenous development. Chiku Malunga is a development management specialist from Malawi, with a special expertise in African traditions as they apply to development. Susan Holcombe, a US national now teaching sustainable development, has lived and worked in Africa and Asia, and has been influenced by her experiences with endogenous development and poverty reduction in China.

Endogenous development is certainly not a mindless rejection of modern ideas, particularly modern science and technology, nor a return to an imagined ideal of indigenous norms and practices in pre-colonial Africa. History cannot go back; it goes forward. Not all traditional norms or knowledge should be carried into the future. We know that all cultures evolve over time and that African traditions and knowledge have and will continue to evolve. What we mean by endogenous development is more complex than a polar opposition of Western and traditional indigenous approaches or philosophies. We are arguing that there is a particular logic to endogenous approaches and endogenous leadership:

- In the histories and cultures of the many societies of Africa we can find many political, social, spiritual, ecological, and economic norms, values, and knowledge that have something to teach, inspire, and facilitate development efforts in the twenty-first century. Over recent centuries, colonialism, slavery, neo-colonialism/natural resource exploitation, poor governance, market penetration, and rapid communications have wreaked havoc with traditional norms, and have sidelined traditional patterns of knowledge generation. Still anthropologists, as well as conversations any of us may have with Africans, can tell the listener that traditional norms about governance, social relations, community, and responsibility remain alive and that traditional knowledge, for example, about management of water use and stewardship of the land, remains active.
- When we think about endogenous development for Africa, we should not just focus on specific political, social, or economic norms, but on the underlying values implied by traditional political, social, or economic institutions and practices. Chiku Malunga, in his article, explores this issue of how we understand traditional indigenous values in a modern era.
- Equally we need to be open to traditional approaches to accountability, water management, and land stewardship that may be overlooked in large-scale development efforts. To the extent that development is driven by external donor funding, there may be few incentives

2

to look at traditional or endogenously developed methods of dispute resolution, water management, and grain storage. Donors are rewarded for introducing innovations, not for taking existing practices and scaling them up.

- Western industrialisation, economic market systems, and social constructs emerged – over time – out of Western history, and the values they represent underpin the 'developed' economies of the North. One can look at the differences between Europe and the US in deeply held beliefs about the functions of the state and responsibility for the marginalised, and observe how these differences may have emerged from historical experience.

- We assume that foreign cultural norms and values cannot be imposed on another society in the name of development (see Fukuyama 2003, 29ff). Our position is that there is an urgent need to recapture and respect indigenous worldviews and wisdom in contemporary development theory and practice, for *there can be no tree without roots*. The values imbedded in endogenous values and wisdom are the roots for growing trees. We remember as well that trees grow and diverge into different branches.

- East Asian countries, with endogenous leadership and building on Asian approaches, have a remarkable record of economic growth, human development, and poverty reduction.

- Twenty-first century African states, with responsible African leadership, can build on their own traditions and values, learning from the West and from other regions, to build the governance, economies, and societies that reflect African priorities and that enhance Africa definitions of the 'developed' society.

- The contemporary development challenges of inequality, livelihoods, responsible governance, and resource stewardship are not new. Throughout the centuries Africans, as individuals and communities, have found ways to address these problems, sometimes with great success. The challenges today may be more acute because of their magnitude, the speed of impacts, the spanning of distances, the deepening of power inequalities, and the growing global awareness of and connection to these challenges.

The experience of peoples from the non-Western world should not be lost but used to enrich our current approaches. This is why we are not arguing for an "either/or" solution but rather a "both/and" solution with more space for locally rooted and driven development practice. This special volume of *Development in Practice* will explore in greater detail the multiple meanings of endogenous development and what endogenous development means in practice. In doing so, it seeks to give voice and access to audience to a variety of African perspectives. Hearing these voices, most from Africa, will not always be easy. Some of the voices in this volume challenge assumptions or values that are central to current development thinking. Think for a moment of the individualism versus the community debate. Development theory and practice today focus on the individual. Human rights approaches are about individual rights, not community rights. Human development and capabilities approaches focus on the individual. Neo-liberal economists insist that economic growth depends on market institutions and on individuals ready to take risks. For example, Acemoglu and Robinson (2012, 262) point to the reciprocating influences of the modernisation of agriculture and a weakening of "*rigid tribal institutions*" (in the late nineteenth century). African farmers' demands for privately owned land weakened tribal chief authority, and individual farmers were ready to adopt technical innovations and increase in the individual income (2012, 261–264).

Bhekinkosi Moyo and **Katiana Ramsamy** plunge right into this debate about the role of individuals versus the community, arguing that African philanthropy (as opposed to philanthropy in Africa) and pan-Africanism have to be the foundation of development success in Africa. In the most basic way, their position is that Africans, like the West before them or the East Asians more recently, have to define the values that underpin African development, and that there is not one Western, cookie cutter model that fits Africa. Moyo and Ramsamy are not rejecting individualism; they are saying instead that Africans must also value community and the African heritage in which it is embedded. African philanthropy, they say, is about *"solidarity"* and reciprocity. The southern African term *ubuntu* captures the underlying values: *"The spirit of ubuntu depicts reciprocity and envelops a communalism of interdependency, sharing, oneness, loving, giving and a sense of a continuum of relationships."* The pan-Africanism that Moyo and Ramsamy discuss is based on principles of African philanthropy and is about the role of African leaders serving their people. Pan-Africanism receded from public view after the time of Nkrumah and others. Nelson Mandela's contribution was to revivify the ideals and practice of African philanthropy and to live these ideals in his own life.

Chiku Malunga takes the discussion of endogenous development to a more concrete level by exploring traditional, indigenous Africa practices in terms of governance, social and community responsibility, inequality, and gender relations. As he outlines traditional norms he goes the next step and identifies the underlying intent and values of traditional practices in the context in which these traditional norms operated. He goes on to suggest how these traditional values can inform development policies and practices in the current context in Africa.

Exploring concepts of endogenous development at the theoretical level is not sufficient. **David Millar**, spurred on by the questions of his students at the University for Development Studies in Northern Ghana, and informed by his years of practical experience,[1] goes from the theoretical definition of endogenous development to a consideration of what this means in practice. Starting with the creation of an enabling environment for endogenous development, he argues that there need to be codes of conduct for development practitioners. Development professionals, whether native or foreign, are usually outsiders to communities where development takes place. They are bearers of information and resources, but they need to be ready to take the time and have the empathy to listen and learn from the cultural context of a particular community. Millar stresses the importance of empathetic relationships.

Charles Banda offers a voice that is usually not heard in a peer-reviewed development journal, even a journal about development in practice. His reflections on the changes in the roles of extended families, of intentional socialisation of adolescent boys and girls, and of intra-family responsibilities suggest that while old customs have eroded they have not necessarily been replaced by new customs that play the same socialisation role in a modern context. This void has left communities particularly vulnerable to modern diseases like HIV/AIDS. For example, boys entering adolescence in Malawi typically lived in boys' communal housing and were instructed by *"respected uncles"* in proper behaviours of cleanliness, sexuality, and family responsibility. The old mechanisms seen narrowly, such as boys' communal houses, may no longer be the right way to prepare young adolescents for life. But the notion of community responsibility for shaping the values of young people could be recreated in more modern variations on the old. All humans have been affected by globalisation and rapid change. For developing countries, the tragedy of globalisation and rapid change is that there has not been time for new cultural institutions to evolve from traditional practice. Malawi, with a high incidence of HIV and a high rate of HIV orphans, represents a case where traditions may be an acceptable, and cost effective, starting point for developing locally rooted approaches to solving a modern challenge.

Endogenous values and approaches in Africa may have remained small scale simply because African have lacked access to (and control over) the means of communicating with other Africans and to non-Africans about endogenous concepts, values, and priorities. While African researchers, scholars, and thinkers have over many decades engaged in research and writing on the African foundations for development and change, little of this work gets into widely shared academic publications or in the popular media. Communications media, including academic journals, as well as disciplinary associations, reside predominantly in the North. Criteria for publishing in journals or for presenting at conferences are set by the North. Southern scholars may lack networks and funding for research or travel, or even for access to ejournals, access which Northern academics take for granted. The barriers to communicating with Africans in the next country or to scholars and practitioners around the world are high. **Dawn Booker**, in "Wiki approaches to wicked problems" suggests that Africa can leapfrog technologies and rely on the wiki collaborative approach and Internet technology to manage the content of endogenous development and facilitate rapid communication and interconnectedness in developing solutions. Booker links wiki approaches to the concept of *ubuntu*, which appears to have originated in southern Africa but is now increasingly used to convey multiple meanings of African humanism, particularly that individuals have identity only in relation to others, to community.

Endogenous development in practice

There are multiple understandings of endogenous development, most of which hold in common that development needs to be rooted in African values and philosophies, and that it needs to be owned and led by Africans. That is a broad understanding that can lead to many interpretations. This section of the special issue presents a range of examples of endogenous development. Rather than starting from a fixed principle of what encompasses endogenous development, the editors want to use an inductive approach and allow readers to examine different examples of endogenous development practice in real circumstances. What follow here are practice notes and research reports that illustrate both different definitions of endogenous development and the complex ways in which endogenous knowledge, awareness of endogenous behaviours and values, and endogenous ownership, participation, and leadership influence development outcomes. The examples reinforce our earlier contention that endogenous development is not a substitute for Western theories of development, but a critical complement that enables local sustainability.

Angélique Rwiyereka's article, "Using Rwandan traditions to strengthen programme and policy implementation", is a direct illustration of how pre-colonial practices of performance goals and accountability have been translated into modern practice in Rwandan Government practice. Implicit in the Rwandan experience is the leadership role of the President and top government figures. *Imihigo*, and the accompanying national and local events of *umwiherero* and *umushikirano*, are associated with Rwanda's success in achieving rapid economic growth and also social development progress. The use of *imihigo* in performance management in Rwanda suggests the need for long-term study of how this contributes to changing the culture of administration in Rwanda, and what effect it might have on knitting together a society previously torn apart by violence.

The ability to manage conflict and create a stable environment for development is a first priority. **Sylvester Galaa** and **Francis Issahaku Malongza Bukari**, both from the University for Development Studies in Northern Ghana, examine the integration of traditional, indigenous participatory processes into the management of tariff payments for a scarce resource – water. Communities in Northern Ghana traditionally regard water as a specific public good, a "gift of nature". The economist might argue that community members need incentives to pay for something that was previously free. In theory the incentive for an attitude change about paying tariffs should be access to safe water and the resulting health benefits for families. Galaa and Bukari found a far more complex situation in the community of Dalun. Neither modern nor traditional conflict resolutions methods worked alone, but results (payment of tariffs) were better when both methods were used. Interestingly they found that traditional cultural modalities, such as songs, dance, and spiritualism, did not have an impact. Willingness to pay water tariffs was not complete and the authors note that changes in attitudes and behaviours around paying water tariffs could not be separated from the capacity to pay in a poor community.

Conflict resolution mechanisms exist in many societies but as **Chimwemwe A.P.S. Msukwa** and **Marion Keim-Lees** point out, there is a critique that traditional *"peace and justice systems [have tended] to exclude women"*. The authors argue that the reality on the ground is more complex and varied by community or ethnic group, and that there is potential in most cases to build on local traditions to build a modern approach to conflict resolution that involves women. They look at patrilineal and matrilineal societies in Malawi, noting that where changes occur in gender relations in conflict management, it is because contemporary practices have built upon a thorough analysis of gender relations and use the gender strengths in tradition as a foundation.

Nathalie Tinguery's practical note explores the intersection between international donor agency projects and endogenous approaches to development. It highlights a phenomenon increasingly seen in INGO field offices where well-educated nationals are increasingly playing leadership roles at the field level and, as an ideal, are bridging the gap between donor strategies and processes and contextual variables that affect the introduction of change. The note suggests that the participatory, bottom-up approach of the donor can move beyond theory and become reality when knowledgeable and sensitive local staff are given the autonomy to implement change that becomes owned by local organisations and communities.

The role of individuals in the diaspora being able to contribute to change and development is being explored in several contexts in the development literature (see Brinkerhoff 2009). Individuals in the diaspora may be able to bridge modern technical change and traditional societies because of their knowledge of endogenous culture, values and social/political structures. **Ariel Delaney** explores the specific case of ADESO in Somalia, founded by Fatima Jibrell, a Somali-American environmental activist. She describes the ways in which ADESO challenges communities to change while relying on traditional culture and values, a belief in people's capabilities, community ownership of change, and the credibility of ADESO.

Professional competence is also expanding in the area of African-led innovation. **Danielle Fuller** and **Kolawole Adeyayo** describe a technical innovation in managing cassava waste that increases income for cassava growers and for goat producers to whom the processed waste is sold – and additionally has environmental benefits. This is an innovation developed endogenously by rural development specialists at the University of Agriculture in Abeokuta in Nigeria. Implementing, testing, and proving the effectiveness of this innovation was made possible by a grant from the World Bank Development Marketplace initiative to fund projects that pilot innovations. This is a story with a happy ending – at least in the short run. Funding was available from another World Bank project and another donor to begin scaling this innovation. The article addresses the challenges that endogenous innovators face, even when they have demonstrated the effectiveness of an innovation.

Endogenous or mother tongue languages are accessible to all within a traditional grouping and are receptacles for endogenous culture. At the same time, some observers argue that colonial languages have the potential to connect different ethnic groups with each other and with the larger world. **Mariama Khan** argues that the use of colonial languages for government business, the courts, schooling, and even agricultural extension may exclude the large percentage of people in African countries who do not speak the colonial language, contributing to inequality and bifurcated societies. While acknowledging the role of international languages, Khan argues that education in the mother tongue makes learning more effective, enhances understanding of cultural heritage, and builds a self-confidence in national capacity. Using examples, particularly from West Africa, she argues that local languages are linked to each other and that people grow up with a facility in more than one local language that enables them to engage in trade and other interchanges. Again, using an example, she argues that while governments may sometimes give lip service to learning in the mother tongue, they have never given priority to the implementation of such programmes. Using mother tongues for schooling and for basic government services does not mean that people cannot have access to international languages.

Donors and endogenous development

While development assistance is not development, the rhetoric and practices of major donors strongly influence development practice, and the funding can be important to development and change. Co-editor **Susan Holcombe** argues that, despite rhetoric about ownership and participation, much donor-financed development remains exogenous. Paradoxically donor priorities for measureable results often have the unintended consequence of inhibiting participation and ownership. Where exogenous actors play heavy roles in implementation in order to produce results for funders, there may be reduced roles for endogenous actors and missed opportunities to build capacity. Holcombe notes the barriers to a greater role for endogenous actors in Africa, most particularly this perverse incentive structure that reinforces top-down approaches, and the challenges of governance. Within the reality of those constraints she identifies some ways to bridge the needs of donors with endogenous approaches that build sustainable capacity and results.

Afia Zakiya continues an assessment of top-down donor approaches that leave little room for understanding and drawing on endogenous experience and history. She writes from the perspective of a Northern NGO engaged in water, sanitation, and hygiene (WASH) projects in Africa and combines both theoretical analysis and observations of practice, particularly the endogenous development approach of WaterAid Ghana. The science of the relationship between safe water and sanitation on the one hand and healthy outcomes on the other is not in dispute. What is missing for successful implementation and sustainability of technical and infrastructure improvements is an understanding of the cultural context and the values, norms, and spirituality that underpin social and behavioural change. The article also suggests changes in the kinds of outputs we measure, going beyond the measurement of numbers of water or sanitation facilities constructed, to a greater focus on tracking behaviour changes.

Conclusion

The articles in this volume represent disparate threads in the existing dialogue around what endogenous development means and how it can be implemented. We do not pretend to have all the answers, though the editors do draw some conclusions in the final article. Most importantly, the experience of putting this volume together leads us to argue for the need to provide space for more authentic voices from the South (including those Southern spaces that exist in wealthy Northern states). If the development community is to hear these voices, it needs to be able to listen even when the voices come in different forms than those to which professional and academic development communities are accustomed. This does not mean the abandonment of science. It does mean an examination of our own biases, and a willingness to engage as equals in a dialogue.

Note

1. He uses the examples of AZTREC and CECIK, and his participation in Compas.

References

Acemoglu, D., and J. Robinson. 2012. *Why Nations Fail: The Origins of Power, Prosperity and Poverty.* New York: Random House.

Brinkerhoff, J. 2009. *Digital Diasporas: Identity and Transnational Engagement.* Cambridge: Cambridge University Press.

Identifying and understanding African norms and values that support endogenous development in Africa

Chiku Malunga

Contemporary development issues are not new. All groups of people, based on their worldviews and contexts, found ways of addressing these societal problems. By their nature, solutions were relevant, legitimate, and sustainable in their contexts. A prerequisite for effective development practice is to understand and respect the roots of African culture. There needs to be a "rootedness" to change and development. Exogenous ideas and practices of potential benefit to Africa must build from the inside out, not outside-in, as an imposition. This article illustrates how African societies have viewed and dealt with these socio-political issues from within.

Les questions de développement modernes ne sont guère nouvelles. Tous les groupes de personnes, sur la base de leurs conceptions du monde et contextes respectifs, ont trouvé des moyens de remédier à ces problèmes sociétaux. Les solutions ont été de nature pertinente, légitime et durable dans leurs contextes respectifs. Une condition préalable pour des pratiques efficaces de développement est la compréhension et le respect des racines de la culture africaine. Les changements et le développement doivent avoir une dimension d'« enracinement ». Les idées et pratiques exogènes pouvant profiter à l'Afrique doivent se développer depuis l'intérieur, et non depuis l'extérieur, comme quelque chose d'imposé. Cet article illustre la manière dont les sociétés africaines ont perçu et géré ces questions sociopolitiques depuis l'intérieur.

Las actuales cuestiones centrales del desarrollo no son nuevas. En este sentido, todos los grupos humanos encuentran maneras de abordar los problemas sociales a partir de sus propias cosmovisiones y de sus contextos y, precisamente por su naturaleza local, estas soluciones resultan relevantes, legítimas y sostenibles en dicho contexto. De manera que, para impulsar prácticas de desarrollo efectivas, es necesario comprender y respetar las raíces de la cultura africana, esto es, la transformación y el desarrollo deben encontrarse «arraigadas» localmente. Así, las ideas y las prácticas exógenas que puedan resultar potencialmente beneficiosas para África deben ser construidas de dentro hacia fuera y no al revés, pues operarían como imposiciones. El presente artículo examina cómo las sociedades africanas han visto y han abordado las cuestiones sociopolíticas desde adentro.

Introduction: endogeneity and development in pre-colonial African societies

Contemporary development issues – governance, hunger, poverty, rights, inclusion, and so on – are not new. They have always existed. All groups of people, based on their worldviews and contexts, came up with ways of addressing these types of societal problems (Bujo 1998). By their very nature, solutions were relevant, legitimate, and sustainable in their contexts. A prerequisite for effective development practice is therefore to understand and respect the roots of African culture. There needs to be a "rootedness" to change and development. Exogenous ideas and practices of potential benefit to Africa must build from the inside-out as a starting point, not be imposed from the outside-in (Wilkinson-Maposa and Fowler 2009). Drawing on the works of Mutwa (1998), Bujo (1998), Fiedler (1996), and Biko (1978), among others, this article identifies and discusses specific examples of values and norms used by traditional African societies to address social and community problems, which have the potential to serve as an endogenous foundation for norms that will allow Africa *today* to address its social and political challenges from within. This article does not discuss negative traditions that contributed to abuses of power, insecurity, exclusion, or loss of rights. As in Europe, Asia, and other regions, not all of our African endogenous norms should be preserved; but, like those in other regions, we Africans need to build on what is positive within our own history. Often, colleagues from Europe have remarked among themselves that "Africa has no history except for the history of Europeans in Africa". This makes them challenge any claims of the civilisation in Africa before and apart from the Europeans. I do not mean to say that the "African way" was perfect – we are talking about ideals that had understandably a good share of shadows and contradictions, just as our modern democracies also have shadows and contradictions.

This article is makes generalisations about endogenous norms and values; the author recognises that sub-Saharan Africa is a vast continent and there are, of course, considerable variations in endogenous practices across Africa (Gyeke 1996, xiv). The following examples are chosen to illustrate how specific traditional practices can productively inform development in Africa. They offer endogenous examples of how norms and practices can have relevance to building good governance, assuring livelihoods and inclusion, and promoting equity and rights.

Governance: the roots of democracy in traditional Africa

Today, the most common leadership style associated with Africa is authoritarian, encased in democratic forms – forms inherited from colonial powers. Efforts by the West to promote democracy are seen as a way to combat authoritarianism; the West looks at elections as the measure of democracy, and elections are thus often enshrined in the practice of authoritarian governments in Africa.

A proper analysis of the original political systems in Africa tells a completely different story of democratic notions. It is a story that is not focused on elections, but on the responsibilities and accountabilities of the good ruler and on the limits to the power of the ruler. In pre-colonial Africa, the king was often not an absolute ruler – that is, he was not ruling alone or without any accountability mechanisms. The king ruled with the help of a council of elders. The king could not make decisions single-handedly; he always had to consult the council of elders. From my experience of working across the continent, this seems to have been a common practice among most African societies (Mutwa 1998). Bujo (1998, 158–161) describes a model of the council of elders from the Barundi people of Burundi.

The composition of the council of elders was crucial. The council consisted of wise men and women. They lived among the people; they were known and recognised by the people; they shared the same living conditions as the people. They had to be people of higher moral standing in society; integrity was a key qualification. Wisdom and integrity were the main considerations in considering one for eldership. Because they believed that wisdom and integrity take time to develop, it was usually the elderly who were considered for these positions. Up to the present, in some traditional societies, one may not be considered for eldership until he or she reaches the age of 60. But there were cases where much younger individuals, who distinguished themselves, could become elders, as the proverb, *"the child who washes his hands will eat with kings"* attests. Class as it is understood today did not exist in traditional Africa. Groups of people were only categorised by their contribution to the community or their occupations, e.g. hunters, traders, farmers, rain-makers, etc. In matrilineal societies, women had more direct space in the composition of the council of elders, while in patrilineal societies ways and means were sometimes created to ensure adequate participation of the women.

This is in great contrast to the current practice in much of the continent. Members of Parliament (MPs) are hardly known by their constituents. Most of them live in far-away urban areas, in affluence. They will only reappear in the village to campaign for the next elections. MPs do not share the same living conditions with the people they claim to represent. They live in great opulence in towns, while their people are living in "hell" in the rural areas. This is an example of outside-in development. Parliamentary forms were bequeathed to Africa at the end of the colonial period, but the forms are not rooted in African culture. As imposed, external forms without adaptation to African culture, parliamentary forms are empty of the informal institutions of integrity, wisdom, accountability, and equality that were implicit in tradition.

Integrity as a key qualification for leadership has become virtually extinct. Today it is the person with the most persuasive rhetoric and with the most money who wins an election. In some countries parliaments are rapidly being taken over by young people from towns with money. Wisdom and maturity are not given the same weight as before. In contrast to the best of indigenous leaders, modern leaders are expected, themselves and through their relatives and cronies, to be *rich*, not to be men and women of integrity. They are expected to give big gifts whenever they visit people, and they are not expected to disclose the source of the money they give as presents.

In the traditional African setting, a member of the council of elders was appointed to be an advisor to the king. When this worked well, he or she was to bring to the king all the problems the people were facing. What is common today tends toward sycophancy and bootlicking, rather than the actual representation of the people's issues. In many African countries, presidents hold almost absolute power and ministers are expected to toe the president's line or risk political excommunication.

A member of the council of elders was expected to act as a just referee on controversial matters. In times of conflict, it was his or her duty to ensure the restoration of peace. A member of such a council was regarded as a father or mother of the people whom they represented. He or she was also a father or mother even of those he or she did not directly represent. Today, most African MPs rarely represent anyone but themselves. If they represent someone, it will usually be their relatives or cronies.

To qualify to be a member of the council of elders one was supposed to undergo a lengthy preparatory and probation period. One had to demonstrate an outstanding level of truth and justice. He or she had to demonstrate a high sense of responsibility, courage, and maturity. In most of Africa today there are no schools to prepare individuals for political leadership roles. Truth and justice are not leadership requirements. One politician in South Africa made a bold statement to the effect that all politicians are liars. Though he was sharply criticised, many people secretly agreed with him, based on their observation of political practice in Africa (Malunga 2010). Most MPs know that they can only demonstrate true courage at the cost of losing everything and being ruined by their political party leaders.

Traditional African ideals set a high standard for leadership behaviour. Bujo (1998, 159) noted that the institution of the council of elders was a serious undertaking, sealed by a kind of consecration. After inauguration into office, a pact was sealed between the elder and the people, requiring the elder to be "related" and therefore accountable to every member of the community. He or she was now expected to care for *everyone*'s well-being at all times. He or she was expected to ensure that the economic and political processes served everyone in the community. When the council of elders served the community well there was no need for a police force. Today, the relationship between citizens and their leaders is characterised more by suspicion and disillusionment. Leaders are increasingly seen as serving their own agenda and interests rather than those of the people.

In addition to demonstrating the requisite characteristics and competences before being admitted to the council of elders, one could be fired from the council of elders if he or she *stopped* exhibiting these characteristics. It was not only achievement that kept the elders in office – it was character that mattered more. Integrity, wisdom, and maturity were given the highest value.

Spirituality also played a key role. It was recognised that the King and elders were ultimately accountable to (a) God – a God of justice and One who is concerned about the welfare of all the people. They took their role as one delegated by and from God. The king was expected to channel God's life-force in the whole kingdom. Any lack of integrity on his part would undermine and diminish this life-force. The high priest played a very significant role as the king's spiritual advisor and as a mediator between the physical and the spiritual. The recognition of the need for spiritual and physical harmony and alignment is not given the same degree of significance today. In many African countries men and women of God have left their higher moral ground and submitted to politicians, and have become their puppets. The prophetic voice of the church has been compromised. It is important to note that the African God, at the level of revelation, was the same in essence as the Christian or Muslim God, introduced into the continent by missionaries. The only difference was that the God introduced by the missionaries came with religions and doctrines which in many aspects were different from the African religion. Both the African and introduced religions believed in God the Creator, the God of justice and integrity and the sustainer of all forms of life. They differed in specific doctrines and beliefs, however, e.g., the need for a specific building in which to worship God, beliefs in a heaven and hell, in the afterlife and denominationalism, among others. In short, the essence of the African belief was spirituality as opposed to organised religion. This spirituality was not in opposition to having material possessions; rather it was about having a proper balance in the attention one pays to acquiring material possessions and ensuring one's spiritual health. In other words, it was about cultivating humane values, a key one being ensuring one's health *so that* one can be of service and use to others.

To take Christianity as an example, when it started in Israel, it embraced the Jewish culture. When it was introduced in Europe, it took on a European culture. When it was transferred to America, it took on an American culture. But when Christianity came to Africa, it largely maintained its European and American cultures and did not build on local cultures and spirituality; rather, in many instances, it sought to destroy these. Bujo (1998) and Fiedler (1996) observed that Christianity would have been more entrenched and had greater impact if the missionaries had sought to understand local culture and spirituality. They argue, for example, that there is a high resonance between traditional African concepts of ancestral worship and Christ: ancestors are the dead who are believed to stand between a clan and God. They take the clan's petitions to God and in turn take God's messages to the clan. The Christ of Christianity essentially plays the same role and fits the description of an ancestor very well: a person who has actually lived and died, and then stands between God and the people. The only difference is that Christ extends the concept of an ancestor from a clan to the whole universe, making God not only a God of specific clans but of the whole universe. This is an important observation and could be seen as a root for efforts to move away from clan or tribal loyalties to national loyalties. If the missionaries had introduced Christianity this way, it would have played a bigger role in lifting loyalties from tribes and clans to national and international levels. Tribal and ethnic conflicts, still common in many African countries, might not be as entrenched as they are today.

The highest expectation of the king was for him to be exemplary. In many ethnic communities the king who was not exemplary could be relieved of his duties by the council of elders and replaced by a new one. It is almost unimaginable in Africa today to ask a sitting president to step down and that he or she would actually do so.

This abbreviated story of the legacy of traditional governance tells us that historically Africans had ideals and traditions of respect for wisdom and experience; of responsible leadership and accountability; and of inclusion and participation. Within traditional African societies there were formal and informal institutions that helped to enforce this ideal of good governance. Responsible leadership, accountability, and participatory governance are values endogenous to the African experience and, despite the impacts of colonial history, independence, and globalisation, these values still resonate with Africans. These are the values on which post-colonial institutions and structures should be based. The challenge is how to do that. Two examples of how traditional values and practices can be used to build modern governance systems from the inside out come from Botswana and Ghana.

In Botswana, the government consciously and proactively promotes the values of *botho* – a local version of *ubuntu* discussed below. According to Adamolekun (1999), traditional Tswana values played a key role in the formulation of the Botswana constitution; ministers and MPs regularly meet with traditional leaders to discuss policy and governance issues. In addition, there is a strong culture of citizen participation through successive competitive, free, and fair elections to local councils and land boards with the aim of strengthening and broadening political participation and enhancing accountability.

In Ghana, the government established the National and Regional Houses of Chiefs and Traditional Councils. Traditional leaders are powerful leaders alongside government leaders, especially in local government. Individuals who have been successful in their careers and professions are becoming traditional leaders. Medical doctors, lawyers, professors, engineers, and successful business people are becoming traditional leaders and bringing with them competences enabling them to mobilise their people around more effective development initiatives (Asamoah 2012; Sakyi 2000).

Equity and inclusion: socio-economic life in traditional African society

Social equity and community responsibility are rooted in African societies. The concept of class as understood in the Western sense did not exist in African societies and communities (Bujo 1998). The African society was egalitarian, recognising people for their expertise and contribution to society. People would therefore be recognised as healers, hunters, food producers, traders, warriors, blacksmiths. Wealth and assets were seen from the perspective of the community, not the individual. For example, property was collectively owned, but traditionally there was sufficient land for each household to make a living. This communitarian thinking and collective ownership of property was not the same as a communist model that distributed benefits equally.

It also differs from the Western model of capitalism. Private ownership is the mainstay of capitalism. Despite its benefits to those that *have*, unregulated private ownership is a cause of much misery in the world, especially among the poor. Modern capitalism has been thriving on transfer of wealth from the poor at the bottom to a few rich at the top. We can see this in the rising Gini coefficients and other measures of income and asset inequality in the US, and also in China as it becomes more market-oriented. In contrast, countries like Brazil, which have started to regulate the flow of benefits to the poor, through programmes such as the Bolsa Familia, have begun to reverse the growth of high income and asset inequality. In Africa we can see the success of cash transfer programmes in Namibia, not simply as a model of state-directed social welfare, but as the modernisation of traditional community responsibility for weaker or marginalised members of society. Africans can give capitalism a human face, and should do so as an evolution of traditional African values from narrow clan or community responsibility to national responsibility. Already, there are many cries calling for a revisiting, and probably a revising, of the whole concept of capitalism.

In the African community model, the household's labour and level of contribution were recognised in sharing benefits. Every member of the community was expected to make a contribution to society, according to their ability. For this reason, every member of society participated in the economy except for very young children and the very old. Everyone who could work worked, but those who could not work were taken care of. Parasitism, free-riding, and exploitation did not exist or were greatly discouraged. To qualify for manhood or womanhood, one had to demonstrate the shared values of responsibility, good behaviour, and hard work. Being left out from one's age cohort, when they were qualifying for adulthood and the recognition that went with it, was one of the worst punishments imaginable. The myriad folktales of the well-behaved, most hard-working and skilled young man who may not have come from the royal line but is rewarded by marrying the king's most beautiful daughter are part of evidence that the contributions of particularly hard or skilled work were recognised.

One of the critiques of this tradition of community and family sense of responsibility in Africa is that it encourages dependency and parasitism. For example, a Zambian newspaper recently noted,

> " … most Zambians are used to dependency. It is part of our culture. The extended family system is a very good social safety net but it has also contributed to the pervasive notion that, if I fail, some relative somewhere will bail me out at no cost." (The Post, "On parasites, debt and the economy", July 5, 2012)

This is a misrepresentation of the indigenous African culture. As noted elsewhere, if one looks beyond the traditional obligation of an income earner to support a family member in need, there is a corresponding traditional obligation of each member of the community to contribute according to his or her ability.

In today's conditions – with stagnating agricultural economies, increasing rural to urban migration, and growing populations – we see the reality that the opportunities for many members of society, particularly young people, have been shrinking. New agricultural methods may decrease the need for labour. Industrial employment has not grown. There are high rates of unemployment and underemployment among African youth. As a consequence, we have large numbers of unemployed youth who have become dependent on family or society or, worse yet, fall prey to risky social behaviours that can lead to HIV/AIDS. The appearance is of dependency and parasitism, but the underlying cause is a failure to provide the means for young people to become productive, contributing members of society. In the case of HIV/AIDS, for example, distributing condoms alone without dealing with the issue of poverty will not solve the problem. Neither will the now-famous ABC (abstain, be faithful, and use condoms) message be enough. When people cannot put food on the table, when they are not responsible contributors, they may engage in risky behaviours just to survive. While acknowledging the important and often indispensable role that condoms play in family planning and in the fight against HIV/AIDS, the indiscriminate distribution of condoms to youngsters is now contributing towards the elimination of African culture. I went to a youth library in Mozambique where there were more condoms than books. Though the use of condoms reduces rates of transmission, the emphasis on condoms as the solution ignores the underlying causes of the HIV/AIDS epidemic, and at the same time undermines traditional norms and practices around a larger responsibility to contribute to the community.

Hospitality was a key characteristic of African societies. It remains so now. The statement, *"don't talk to strangers"*, does not exist in Africa. One did not need to know you to be kind to you and to accommodate you. This is why the first white settlers did not meet much resistance from the locals; unfortunately and self-servingly, they mistook the locals' hospitality as a sign of weakness.

Hospitality was also linked to generosity. I remember when I was growing up that whenever my mother cooked a chicken, which was considered a delicacy at that time, she would go around the compound giving the neighbours a piece of the chicken before we could eat ourselves. I imagine if she did the same today, she would be looked at with a lot of suspicion and even ridicule. Hospitality is a tradition to be nourished as a strength. In modern Africa it can be seen as a national and even global value where one respects and honours strangers, creating a basis for a national identity within the colonially-set borders of Africa. The hospitality industry in our countries can derive its spirit from the traditional values of hospitality. Most times, we fail to make a conscious link between the modern hospitality industry and traditional hospitality values.

Tradition can also be used to address one of the most poignant problems of Africa: those orphaned by HIV/AIDS. In some countries these orphans represent 10% of the total child population. Traditionally, orphans did not exist in Africa. Children belonged to all and parenting was not the sole responsibility of the biological parents; parenting was collective. Every grown-up was responsible for parenting all the children in the community. Any elder could mete discipline to a child without having to get permission from the child's biological parents. Traditionally, this is seen as how many African societies deal with the issue of orphans. Ideally orphans are always taken care of by close relatives and the whole community. In some communities where HIV/AIDS is rampant, the capacity of the community to care for orphans is diminished by illness and deaths. The value, that children should always be nurtured by their own communities, can still be maintained. In the case of communities decimated by disease, local governments (and working with them, donors) can provide appropriate support so that orphans can be raised in the community, going to school and getting the support of elders and all community members in preparation for life. There are examples in Malawi and Kenya, for example, of this kind of government and other support for keeping HIV/AIDS orphans in their home villages.

Increased life expectancy is one of the benefits of improved healthcare. Old people's homes and isolation of the elderly did not exist in traditional Africa; parents were taken care of by their children until they died. The values of traditional African society maintained that people contributed to the extent that they could and were taken care of when they could not. This practice is still valued by most Africans, but the exigencies of modern life, including migration to cities and urban housing, make caring for the elderly more difficult than when the family resided in one village. Families, communities, and policymakers need a dialogue about how to retain this healthy tradition of care for the elderly, and what adaptations need to be made for the current and evolving context.

Funeral management companies are a new introduction in most of Africa and in many places they are facing stiff resistance as Africans feel they are taking away their cultural space. The same is true for wedding management companies. Funeral services and rituals play a key role in building community relationships because in the African tradition, no matter where one is, one must travel to bury relatives and friends. It is also during funerals that traditional leaders find space and time to raise key issues facing the community. Similarly, weddings are not just about two individuals marrying, but about two families becoming one, and the wedding is a way of formalising this relationship. Hiring a consultant to handle this process defeats this purpose. The developmental value of this is the obligation it gives the couple to commit to not only each other but the wider group involved. This strengthens both marriage and community loyalty.

Traditional practices around childbirth, initiation rites, marriages, and burial ceremonies were all carried out in ways that involved everyone in the community and required a lot of time. In new, increasingly urbanised settings, the practices may be changed somewhat, but the principles behind the practices need to be preserved at all costs. In many African communities today the mourning period for the dead, for example, has been drastically reduced from months to a few days because of the increased demands on time in the modern context.

Gender

In traditional Africa, men and women were related mostly through the concept of life: the greatest command for individuals and the community as a whole was the promotion of life. By promoting life, the individual enhances the life force of the whole community; by constraining life, one diminishes the whole community. Children were seen as the greatest and highest gift to the community. Both men and women were involved in this process of giving and promoting life. From this point of departure, the relationship between man and woman was not one of superiority or inferiority but one of complementarity of roles to the benefit of the community. The male was never complete without the female, and vice versa.

According to Bujo (1998, 123–126) the roles of the male and the female were divided as follows:

The male role

The primary role of the man was to engender life and support the woman during her pregnancy and in taking care of the children until they were grown up. The role of the man was to support his wife until his death. A number of taboos regulated the support that the man gave to the wife to ensure continued health and harmony. A number of taboos regulated the man's sexual relationship with the wife from pregnancy until the baby was born, and from that time until the child stopped breastfeeding. This understanding of the role of the man is contrary to much of what is normally believed and understood about the male role in traditional Africa. It is believed by many that in Africa, the man leaves all the hard work to the woman. This is a misunderstanding and misrepresentation of African culture. Where men left gardening or farming to women, it means that they were engaged in work that was relatively much harder than gardening. Men have traditionally been involved not only in gardening but also in animal rearing, hunting, and defence of the community. The processes of colonisation and later globalisation brought changes to some of these roles over the past 200 years. In British colonial Africa, males had to work for cash in order to pay taxes, or maintain roads. Later, men were recruited to work in mines in South Africa or urban industries. Then development projects introduced commercial crop production for export, employing men not women. As a result, men were away from the farmstead and women took on greater roles in production on the farm and outside the house. Today, urban women are increasingly working outside the household. The important point may be that we need to be careful about making judgements when we look at household surveys that document the greater number of hours women work. It does not mean that men are necessarily lazy; it may reflect the negative influences on gender roles of colonial extraction, slavery in West Africa or the East African coast, urbanisation, migration, commercial farming, and also the imposition of Western gender values on urbanising African society. Colonial societies in Africa were highly male-dominated, and gender equity is a relatively recent phenomenon in the West.

Traditionally, inherent values and attitudes of African men towards women were responsibility, respect, and care. Most modern gender interventions are oblivious to this fact and start from the point of view that these values must be introduced into the community rather than retrieved from people's cultural values. Progress in gender interventions, especially among men, would be faster if men were made to understand that more equitable gender roles are inherent to their cultures and traditional values, rather than a foreign concept being introduced from outside.

The female role

The woman was the giver of life. She gave birth to children, ensuring the sustainability of humanity. Through her motherhood role she was the primary teacher of all human beings. She played a key role in turning babies and young children into full grown-ups. The mother carried the baby in her womb and then at her back until the child began to become independent. In this way, the baby was gradually prepared for adult life. By initiating life in her womb, the woman was much more connected to the mystery of life and to God, who was the source of all life.

Being the one who gave birth, the nature of the woman was perceived to be directed inwards while in relative terms the nature of the man was perceived to be directed more outwardly. The man was therefore dependent on the woman as far as issues of wisdom and values are concerned, while the woman was dependent on the man for external issues. This emphasises the principle of complementarity of man and woman for completeness of life.

The woman played a key social role in relationship building. Through marriage the woman connected two lineages. Women also played key roles in religious aspects of life, especially in matrilineal societies. Women also played key economic roles in terms of production and aspects of trade. In most of today's gender interventions the starting point is that women are victims of the gender relationships between men and women. The conscious appreciation that women *were* equal partners with men, playing different roles in society, would be a good starting point for development interventions. Building on this, debates would form a stronger and more meaningful foundation for gender interventions if they focused on what the evolution of this over time might mean in terms of the necessary future changes in relationships, roles, and responsibilities of women, respecting the principle of partnership between men and women. In addition, cultivating "female" values of relationship-building would play a critical role in most team-building, conflict resolution, and trust-building interventions.

Poverty

The concept of poverty as understood today did not exist in African traditional society. The primary responsibility of the king or queen was to ensure the prosperity of his or her people.

Prosperity was seen as a blessing, a sign of the success of the leadership. Prosperity was based on the productivity of each member of the kingdom. Problems like unemployment did not exist. Prosperity was also based on trade and not aid. Different kingdoms and groups did their trade through barter, one group providing what the other group needed.

Prosperity was understood not only in its material dimension but holistically. The chief priest not only played the role of a prophet but also the role of a strategist, for instance if a catastrophe befell the kingdom, threatening its prosperity. It was his or her responsibility to find out the root causes and solutions through spiritual revelation. They understood that the causes of poverty have just as much to do with spiritual as well as physical sources.

These three points – the role of leader or of leadership ensuring the prosperity of their people; basing prosperity on productivity and trade rather than aid; the importance of taking a holistic approach to understanding and addressing the issue of poverty – are mostly absent in much of the discourse in addressing poverty on the continent today. African poverty today is mostly dealt with as a solely socio-economic issue. For this reason, most of the efforts on the continent are failing – because they do not go deep enough. They do not touch the issues of culture, identity, and spirituality. It is the clarity of the issues of culture, identity, and spirituality that enable a people to determine for themselves what would enable them, to articulate their agenda and the type of development they want – especially in engaging with outside others like donors.

Human rights

Human rights provide common ground for all humanity as a basis for world peace. Most of the global instruments for this are based on the Western worldview. They are derived from an individualistic concept of a human being rather than the communitarian concept of identity, used in Africa for example. The focus is the individual and not the community, which goes against the African worldview.

Two examples illustrate that human rights as they are constructed today are not so universal as they claim to be, so far as the African culture is concerned.

The Universal Declaration of Human Rights in article 26, 3 (quoted in Bujo 1998, 154) says, "*parents have a right to choose the kind of education that shall be given to their children*". In traditional Africa this does not hold because parenthood is understood differently. One's parents are not only the biological parents. One's parents are all the people older than oneself in the community. They are all responsible for the education of all the children in the community. If the children are not educated, not only in the modern Western sense, all the elderly people in the community take the blame. It is the whole clan or community that raises the child, not just the biological parents. To be more relevant to Africa, the Universal Declaration regarding the education of children would have to deal more with community rather than the biological parents or the nuclear family alone. It is also important to note that in modern times community should not be narrowly understood to mean only ones kinsmen or tribe. In this sense there is responsibility of the larger, national community, to provide education for children. There is no contradiction between individual and community rights. Individuals have rights and communities have legitimate expectations from their individuals. Where tensions arise this principle helps moderate the dialogue.

The second example has to do with free consent in marriage. The Universal Declaration of Human Rights Article 16, 2 says, "*marriage should be entered into only with the free and full consent of the intending spouses*". In Africa, marriage is traditionally community- rather than individual-oriented. In the West marriage is a contract. In Africa marriage is a covenant between two communities, including all their living and dead members. While consent of the intending spouses is paramount and often indispensable, there are more stakeholders beyond these two. It is very important to get the blessing and consent of parents from both sides before entering into the marriage. If two people begin to live together as a man and wife without the consent of the parents, chances are high that the marriage will not succeed. Since the marriage is a *community* covenant, the communities are there to support the couple through the difficult times in a marriage. If it so happens that the two people begin to live together without following the expected marriage process, usually the parents from both sides will not rest until the situation is corrected. Marriage is a foundation for society, therefore the principle of marriage as a community covenant, extending *beyond* the two intending partners, helps strengthen community and social cohesion which are key tenets of any development work.

In Africa, it is almost impossible to live like a Western nuclear family without the interference of in-laws and other relatives. In-laws and relatives have a right to meddle as long as they do it reasonably. Being reasonable means that they can meddle as long as they do not cause unnecessary and unmanageable tensions for the spouses. This has to do mostly with financial demands.

Divorce cases are relatively less frequent in traditional Africa because there are many stakeholders to the marriage. The stakeholders create many checkpoints for entering into a marriage, and equally they create many barriers for exiting the marriage. Each marriage has specific individuals appointed to be special counsellors. These are appointed from both the wife's side and the husband's side. The role of the counsellors (*ankhoswe*) is to build the marriage no matter what disagreement or conflict arises. Their role is never to tear apart. In the rare incident of a divorce, the *ankhoswe* would have done everything within their power to ensure the marriage survived. It is the weakening of the institution of *ankhoswe* that can explain some of the increasing cases of divorce, especially among young couples in urban areas.

Environment

One of the most frightening issues facing mankind today is climate change and environmental degradation. Many people believe that we are all headed towards an inevitable oblivion, the main cause being that we are taking more from the planet than can be replenished. The promotion of consumerism to sustain growth goes against the African belief of promoting rather than diminishing life. In Africa, life is the most sacred thing. For an African, the most important word is "relationships" – relationships with fellow human beings, nature, and the spiritual world. Health is connected to harmonious relationships among these; life is enhanced when these relationships are harmonious, and is diminished when these relationships are not harmonious.

For the reasons given above, African culture has great respect for nature and the environment. It is a common belief that one must use from nature and the environment only what one needs. In hunting for example, it was commonly believed that it is not right to kill the healthiest male and female animals as the healthiest were best able to ensure the sustainability of the species. It was believed that any imbalance would eventually destroy the environment and consequently harm life, which was the highest value among the people. Failure to consider the importance of this balance is leading to serious consequences in the world today. One thing is clear: by cutting the next tree which is not replaced we are cutting our own lives. We are that connected with our environment. Today, because of high levels of absolute poverty and the search for short-term solutions, policies are increasingly oblivious to these considerations. Extractive mining of oil and other minerals take priority over environmental safety and health. Large tracts of fresh water are destroyed for the extraction of oil, rivers are destroyed by waste from mines, forests are being wiped away without replacement, and holy sites with cultural and spiritual value are being desecrated in pursuit of money.

Sidetracked

Development is a natural process. Like all living organisms, development is greatly shaped by the natural environment and context. Development is therefore endogenous in nature – it moves from inside outwards and not from outside inwards.

As mentioned above, developing countries were on their natural journey of development until they came into contact with the imposed processes of slavery, colonialism, imperialism, and globalisation. These processes reversed the natural order of development from inside-outwards to outside-inwards. The highway of the natural process of development was taken over by the imported "outside-in" process of development; the natural endogenous process was detoured and marginalised. If considered at all, it was lumped with mostly non-developmental issues. Development was put in the hands of imported development models. Today, endogeneity as a key development factor remains a mostly peripheral issue. A recent massive (348-page) World Bank report, *Localizing Development: Does Participation Work?* (Mansuri and Rao 2013), shows a worrying loss of memory. The study lacks *any* mention or appreciation of endogeneity as a crucial dimension for participation, though it was a critical feature of the Bank's sensitive approach to context and culture some 20 years ago (Serageldin and Tabaroff 1992).

Although indigenous wisdom does not feature much in modern development models, one observes that among the local people this wisdom still features strongly, especially on social issues. Indigenous wisdom is applied to community issues like childbirth, rites of passage, marriages, and deaths and funerals.

A common phenomenon among many developing country professionals is to live in two worlds at the same time: the modern world at work and the traditional world at home. This sometimes gives a feeling of disconnection. True effectiveness is reached with harmonisation of the realities and selves that one is working and living with.

A key developmental challenge today is to take the "natural process of development" away from the Western diversion or detour, back to the main road of development (Chilisa 2012,). This is the only way of ensuring an authentic, home-grown, grounded, and context-driven development. This is the way to claim and gain control of the future for the developing countries by their citizens.

It is important to note that by taking the natural process of development back to the main road we do not mean going back to the good old days or the "re-villagisation" of development. This was one of the key weaknesses of the "building on the indigenous" school of development thinking some 30 years ago. In misinterpreting Claude Ake's (1988) enjoinder, this was taken to mean loading external project management designs and competencies on what was local and, if it could not be found, inducing aid-dependent structures to emerge. This legacy needs to be selectively evaluated. Development must always start with what is already there; then we as Africans need to see how to use this as a foundation for adaptation and development. In this process, the outsider must always play a helping rather than a leading role.

Being rooted does not mean going back to the dark ages, but optimising or leveraging the digital age in which we are living for endogenous development. It is forward- and not backward-looking. It is about making our traditional principles and ideals find relevance in the current and emerging society – not societies that no longer exist. This is the same for the practices. We need to make a distinction between what traditional practices still make sense and which do not. This means preserving what is useful and discarding what is not. In essence, it is more about the relevance of the *principles* rather than preserving traditional practices specifically.

It is important to note that, while much of what has been presented here about endogenous development is positive, there is also a shadow side to it. According to Malunga (2009, 2), some of the shadows include:

- loyalty to kinship developing into tribalism;
- the practice of kings and chiefs ruling for life, leading to modern elected leaders not respecting term limits in office;
- fear of unpredictable futures, motivating leaders to try to accumulate as much wealth as possible or succumb to corruption while in office;
- values attached to relationships at the expense of personal progress often leading to wasteful expenditure on, for example, births, weddings, initiation ceremonies, and burials;
- the value of respect for elders may lead to blind loyalty to old ideas that may have stopped working; and
- the desire for continuity or survival of the village or clan can undermine the need for radical change in response to rapidly changing environments.

This means that effective development practice is one that is conscious enough to maintain what still serves a useful purpose, to jettison what may have stopped working; and to be sensitive enough not to throw away the baby with the bath water.

Life is one-directional, it never goes in reverse. By going back to the main road we mean setting world-class standards of theory and practice which are sensitive to the current realities and complexities in developing countries and the futures to which their peoples aspire. The futures aspired to are not a mere imitation of the current developed countries; neither are they a return to the good old days of a "glorious ancient Africa". Being rooted means the ability to develop visions of dignity, peace, and prosperity that build on the best of who we are, where we are coming from, where we are today, and where we collectively and consciously want to be – the world we want our children to live in. Development efforts that are not rooted or aligned to this vision cannot be truly effective. It also means building on what we admire from the developed countries but also avoiding what we do not like from them.

We will need to build the new model of development on relevant indigenous principles and values; and also to build on the best of what we have learnt from the processes of slavery, colonialism, imperialism, and globalisation which are now in the main road and need to be removed to pave a way for the new model. To paraphrase Fowler (quoted in Obadere 2013), what we need is a deeper appreciation of locally rooted and driven development practice rather than development practice (designed and developed elsewhere) with local characteristics. The essence must be local but with world-class standards. It is naïve to think that going back to our traditional values and practices alone will do. It is also naïve to think that everything developed from and after slavery, colonialism, imperialism, and globalisation must be discarded. *Wisdom is like a baobab tree; no one person can embrace it alone.* This approach will ensure relevance not only to the developing countries alone but to the developed countries as well, making a developmental contribution to both.

Conclusion

There must be a re-examination of the framework of Africa's development, moving away from the one prescribed by the West towards an endogenous one. The starting point in taking the natural process of development to the main road is identifying the framework that guided the understanding and practice of development and capacity building among the indigenous peoples. Different people in different parts of the world had their different frameworks. From an African perspective, development is understood holistically. Relationship is the foundation for the being and doing of development. An African lives in and for the community; the individual cannot exist without the community and the community cannot exist without the individual. The conscious interdependence between the individual and the community is what characterises that which is essentially African. The challenge is to move from the traditional to a modern understanding of community. Traditionally community could mean one's family, clan, village, or tribe. Today, and developmentally speaking, community means all these but most importantly it refers to national, continental, and global belonging and loyalty: community therefore means loyalty to all of humanity. The demand this understanding makes on an individual or a group of people is to identify one's gift or talent and make this gift one's contribution to the betterment of the world. Governments have also obligations to meet the legitimate needs of their citizens. This should be the essence of development practice at all levels.

This framework is a guide to what processes and content must fill development efforts and practice. Its strength lies in building on local values and priorities while being open to relevant progressive outside influences. With appropriate dialogue there need not be contradictions between pursuing individual *and* community rights, as these two necessarily need to feed each other. This is another way of talking about rights and responsibilities. The individual has some rights but at the same time he has some responsibilities towards the community; the community has some rights and it also has some responsibilities towards the individual.

The papers and contributions to this special issue illustrate how this important task of taking endogenous development back to the main road is evolving on the continent. The contributions show the potentials, challenges, and lessons to be learnt in order to ensure successful endogenous development.

References

Adamolekun, L. 1999. "Pragmatic Institutional Design in Botswana – Salient Features and an Assessment." *International Journal of Public Sector Management* 12 (7): 584–603.

Ake, C. 1988. "Sustaining Development on the Indigenous." Paper prepared for the Long-Term Perspectives Study, World Bank, Special Economic Office, Africa Region (SEO AFRCE 0390), Washington, DC, December.

Asamoah, K. 2012. "A Qualitative Study of Chieftaincy and Local Government in Ghana." *Journal of African Studies and Development* 4 (3): 90–95.

Biko, S.1978. *I Write What I Like.* Johannesburg: Picador Africa.

Booth, D. 2012. *Development as a Collective Problem: Addressing the Real Challenges of African Governance.* London: ODI.

Bujo, B. 1998. *The Ethical Dimension of Community: The African Model and Dialogue Between North and South.* Nairobi: Paulines Publications.

Chilisa, B. 2012. *Indigenous Research Methodologies.* Los Angeles, CA: Sage.

Fiedler, K. 1996. *Christianity and African Culture.* Blantyre: CLAIM.

Fowler, A. 2013. "Civil Society and Aid in Africa: A Case of Mistaken Identity." In *The Handbook of Civil Society in Africa*, edited by E. Obadere, 417–438. New York, NY: Springer.

Gilles, T., and F. Avardo. 2012. *Country Systems Strengthening: Beyond Human and Organizational Capacity Development.* Washington, DC: USAID.

Gyeke, K. 1996. *African Cultural Values.* Accra: Sankofa Publishing Company.

Malunga, C. 2009. *Understanding Organizational Leadership through Ubuntu.* London: Adonis and Abbey Publishers.

Malunga, C. 2010. *Oblivion or Utopia: The Prospects for Africa.* Lanham, MD: University Press of America.

Mansuri, G., and Rao, V. 2013. *Localizing Development: Does Participation Work?.* Washington, DC: The World Bank. Accessed May 28, 2014. http://siteresources.worldbank.org/INTRES/Resources/469232-1321568702932/8273725-1352313091329/PRR_Localizing_Development_full.pdf

Mutwa, C. 1998. *Indaba, My Children: African Tribal History, Legends, Customs and Religious Beliefs.* Edinburgh: Canongate.

Sakyi-Kojo, E. 2000. "Gone But Not Forgotten: Chieftaincy, Accountability and the State in Ghana 1993–1999." *Africa Sociological Review* 7 (1): 131–145.

Serageldin, I., and J. Tabaroff. 1992. *Culture and Development in Africa, Vols. 1 and 2*, Proceedings of an International Conference, The World Bank, Washington, DC, 2–3 April.

Wilkinson-Maposa, S., and A. Fowler. 2009. *The Poor Philanthropist II: New Approaches to Sustainable Development.* Cape Town: Centre for Leadership and Public Values, University of Cape Town.

Endogenous development: some issues of concern

David Millar

This article aims to provide knowledge and practical guidance to managing and implementing within the framework of endogenous development. The paper gives a theoretical overview of endogenous development, linked to issues of globalisation and poverty, and ongoing work among European institutions and academics that suggest shifts in Europe from exogenous to endogenous development approaches. It then makes a case for a paradigm shift – an African alternative to modernisation and development, namely endogenous development – using experiences with two NGOs in Ghana and Zimbabwe to locate theory in practice. The paper concludes with some empirical pre-requisites for conducting endogenous development with rural communities.

This article is prompted by the requests of my students at the University for Development Studies, Ghana, for knowledge and information, and practical guidance to managing and implementing within the framework of endogenous development. I start by giving a theoretical overview of the concept of endogenous development and link it with current issues of globalisation and poverty. I briefly mention current work among European institutions and academics that suggest shifts in Europe from exogenous to endogenous development approaches. Encouraged by such developments, I then make a case for a paradigm shift – an African alternative to modernisation and development, endogenous development. I bring to light the experiences with endogenous development in two NGOs – CECIK (Ghana) and AZTREC (Zimbabwe) – in order to locate theory in practice (praxis). I conclude by providing some empirical prerequisites for conducting endogenous development with rural communities, which demonstrate one way of conducting experimentation with farmers within the context of endogenous development.

Cet article a pour but de fournir des connaissances et des conseils pratiques pour la gestion et la mise en œuvre au sein du cadre de développement endogène. Il propose un aperçu théorique du développement endogène, lié aux questions de mondialisation et de pauvreté, et des travaux en cours parmi les institutions et les universitaires européens qui suggèrent une évolution en Europe, d'approches de développement exogènes à des approches endogènes. Il présente ensuite des arguments pour un changement de paradigme – une alternative africaine à la modernisation et au développement, le développement endogène – en utilisant des expériences avec deux ONG au Ghana et au Zimbabwe pour situer la théorie dans la pratique. L'article se conclut par quelques conditions préalables empiriques pour entreprendre le développement endogène avec les communautés rurales.

Cet article est motivé par les demandes émanant de mes étudiants de l'University for Development Studies, du Ghana, désireux d'acquérir des connaissances et des informations, ainsi que des conseils pratiques pour la gestion et la mise en œuvre au sein du cadre de développement endogène. Je présente pour commencer un aperçu théorique du développement endogène et le relie aux questions actuelles de mondialisation et de pauvreté. Je mentionne brièvement les travaux en cours parmi les institutions et les universitaires européens qui suggèrent une évolution en Europe, d'approches de développement exogènes à des approches endogènes. Encouragé par ces progrès, je présente ensuite des arguments pour un changement de paradigme – une alternative africaine à la modernisation et au développement, le développement endogène. Je mets en relief les expériences avec deux ONG – CECIK (Ghana) et AZTREC (Zimbabwe) – afin de situer la théorie dans la pratique. Je conclus par quelques conditions préalables empiriques pour entreprendre le développement endogène avec les communautés rurales, qui illustrent une manière de mener des expériences avec les paysans dans le contexte du développement endogène.

El objetivo del presente artículo se orienta a brindar conocimientos y orientaciones prácticas en torno al manejo y la implementación de acciones en un contexto de desarrollo endógeno. En este sentido, presenta una revisión teórica de dicho desarrollo, vinculado a los fenómenos de la globalización y la pobreza, así como a las investigaciones en curso llevadas a cabo por instituciones y académicos europeos. Tales investigaciones apuntan a la existencia de una transición en Europa desde enfoques de desarrollo exógeno hacia enfoques endógenos. Asimismo, apoyándose en las experiencias de dos ONG de Ghana y de Zimbabue, se plantea la necesidad de un cambio de paradigma, que implique considerar una alternativa africana para llevar adelante la modernización y el desarrollo (el desarrollo endógeno), con el fin de aterrizar la teoría en la práctica. El artículo concluye señalando algunos prerrequisitos empíricos que resultan necesarios para que sea posible realizar el desarrollo endógeno conjuntamente con comunidades rurales.

La preparación de este artículo fue motivada por la insistencia de estudiantes que recibieron cursos del autor en la Universidad para los Estudios de Desarrollo de Ghana. Los mismos deseaban recibir conocimientos, información y una orientación práctica respecto al manejo y a la implementación en el marco del desarrollo endógeno. El artículo comienza realizando una sinopsis teórica del concepto de desarrollo endógeno y analizando la forma en que se vincula con el fenómeno de la globalización y la pobreza. El autor revisa someramente las investigaciones actuales realizadas al respecto por instituciones y académicos de Europa. Dichas investigaciones señalan que el interés que anteriormente se centraba en el desarrollo exógeno actualmente se está desplazando hacia la promoción del desarrollo endógeno. Inspirado por esta transición, el autor postula la necesidad de plantear un cambio de paradigma —el surgimiento de una alternativa africana para la modernización y el desarrollo, es decir, el desarrollo endógeno. Asimismo, con el fin de relacionar la teoría con la práctica (praxis), hace referencia a las experiencias obtenidas en el área de desarrollo endógeno en el seno de dos ONG —CECIK (Ghana) y AZTREC (Zimbabue). El artículo concluye mencionando algunos prerrequisitos empíricos empleados a la hora de impulsar el desarrollo endógeno en comunidades rurales, los cuales dan cuenta de una manera de realizar ensayos conjuntamente con campesinos en el contexto del desarrollo endógeno.

Introduction

When lecturing on endogenous development to my final year BSc Agriculture students in the University for Development Studies in the north of Ghana, I noticed the great interest and enthusiasm, but also the bewilderment, frustrations, and doubts, in these young minds. It was therefore small wonder that after what I thought was a brilliant presentation of the subject I was met with the following questions by some keen students: "*What is this endogenous development all about? How can we differentiate this development approach from the existing ones on indigenous knowledge? How is it also different from existing participatory processes? What practical experiences do we have with endogenous development in Ghana and in Africa? Do you have any writings on them from your own work?*"

I found their questions very pertinent for the entire development arena. My answers to these questions, coupled with the accumulation of theoretical and practical experiences, have been put together here for a wider audience beyond my students at UDS-Ghana. My focus is not so much the answers to the questions my students raised, but my own reflections on the concept of endogenous development. I start with the theoretical foundations of endogenous development and conclude with practical experiences.

Theoretical discourses on endogenous development

"Endogenous development implies development from within that is both biophysical and socio-cultural in nature. It is based mainly, though not exclusively, on locally available resources, local knowledge, culture and leadership, and their cosmovisions, with the openness to integrate outside knowledge and practices." (Haverkort, Millar, and Gonese 2003, 6)

This form of development is more akin to African systems of agricultural productivity than most other previous paradigms such as Transfer of Technology (TOT), Trickle-Down Approach, and even Farmer Field Schools (FFS).

Endogenous development patterns are based mainly, but not exclusively, on locally available resources: the ecology, labour, and knowledge of an area as well as those patterns that have developed locally to link production and consumption. Enhancing endogenous development implies building on local resources and complementing them with appropriate external resources, maximising local control, encouraging the dynamics of local knowledge systems, retaining of benefits within the local area, and ensuring equity in the sharing and use of resources. This process also implies networking, lobbying, and policy advocacy leading to policy reforms (Haverkort and Hiemstra 1999, 12). It is largely a self-oriented growth process, the benefits of which are largely retained within the region of production.

In contrast to this, Van der Ploeg and Saccomandi (1993) and Van der Ploeg and van Dijk (1995), in their studies of European agriculture, have described the tenants of *exogenous development* to include comparatively high levels of transaction costs, high management costs, and less balance between transaction and transformation costs.

Endogenous development in perspective

To deal with the endogenous development I start with the discourse about worldviews and cosmovisions. Through our action research in Asia, Africa, and Latin America, I have come to learn that local knowledges in most cultures include a wide diversity of assumptions, concepts, technologies, and ways of experimenting, teaching, and learning that are specific to the culture and ecosystem.

My work so far has also brought to light that, even with the immense diversity in the ways local knowledge is phrased and expressed, a common feature is represented by conceiving life in terms of three inter-related and inseparable domains:

- the natural world
- the social world
- the spiritual world

Local knowledge in the natural domain includes thematic fields related to specific agricultural, health, and other practices. It includes knowledge about the physical world, ("dead") material, and that about the living world: the biology of plants, animals, and humans.

The social domain includes knowledge about local organisation, local leadership, and management of natural resources, mutual help, conflict resolution, gender relations, art, and language.

The spiritual domain includes knowledge and beliefs about the invisible world, divine beings, spiritual forces, and ancestors, and translates into values and sense-giving and related practices such as rituals and festivals.

An important feature is that none of these domains exist in isolation. In many traditional ways of knowing a notion of unity exists according to which the natural, social, and spiritual worlds are considered to be inseparable and integrated. This gives rives to the concept of *cosmovision of a people*. For the African, Figure 1 is a reconstruction of the cosmovision of the Dagaaba and Gruni of northern Ghana and provides detail of the three domains.

Hence, endogenous development aims at the local determination of development options: local control over the development process and the retention of the benefits of development within the local area. Endogenous development can be looked upon as in situ conservation and development. It takes the local values, local concepts and practices (with their biophysical,

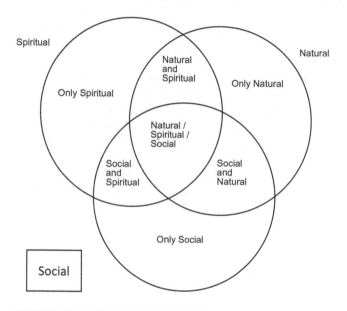

THE 3-CIRCLES DEPICTING THE AFRICAN WORLDVIEWS

Social World, Spiritual World, and Natural World - the interaction of the three worlds implies the following constellations of knowledges:

- Knowledge resulting from Social interactions only.
- Combination between the Social and Natural.
- Combination between the Social and Spiritual.
- Knowledge resulting from Natural interactions only.
- Combination of the Natural and Spiritual.
- Knowledge resulting from Spiritual only.
- Combination of Social, Spiritual, and Natural (*this last constellation is the perfect state which strives to be in balance or harmony with itself*).

Figure 1. Constellations of cosmovision-related knowledge: the three circles depicting African world views.

socio-economic, and spiritual dimensions), local resources, and opportunities as the starting point of development. However, it does not assume that indigenous knowledge will have all the answers to present-day problems.

In this discourse, one is aware that indigenous knowledge and its values may have their limitations. The ability of indigenous knowledge to adapt to present needs can be limited; it may not be uniformly distributed in the community and the individual aptitudes to generate and accumulate the knowledge may be different. In traditional societies access to specialised knowledge may be limited to certain persons and its use does not necessarily benefit the community. Yet, also in these cases, the major spiritually related decisions in these communities are based on the traditional knowledge and values and therefore they need to be addressed in development initiatives. Issues of equity, marginalisation, and other rights tend to water down the significance of such a choice (an example is land and land-related issues). However, these are challenges to deal with in constructive discourse, rather than "throw away the baby with the bath water". Hence I posit that the main challenges of endogenous development are to enhance the dynamics of the local knowledge systems and to identify new development niches based on the comparative advantages of the specific eco-cultural situation.

To paraphrase Haverkort, van't Hooft, and Hiemstra (2003b, 30), endogenous development is thus seen as an approach that takes place alongside the ongoing technological and economic global processes. It has the potential to address local needs and contradictions, use local potentials, build local capacity to organise to take initiatives and negotiate, link local economies to international systems with optimal terms of trade, and allows for the co-existence of different cultural identities. I see it as a structural approach to poverty reduction in marginal areas. Intercultural, research, and policy dialogues are approaches to support endogenous development in a regional, national, or global context. How can one conduct endogenous development processes in Africa?

Globalisation, poverty, and endogenous development

Globalisation offers opportunities to link people across the globe in order for them to exchange information, goods, and services. In the present global information system people can inform and learn from each other, assist each other in decision-making, and join forces in negotiation and lobbying. Globalisation has contributed to fast and intensive communication and greater knowledge about different societies, cultures, and ecosystems in the world. Another result is an increased awareness of the fragility of the earth's ecosystem, the differences in wealth and poverty, and the relevance of cultural diversity.

Globalisation is purported to have contributed to overall economic growth, but this growth is not found everywhere and not all social categories have benefited from it. Many are forced to migrate and traditional life forms are driven into the background. Poverty can still be found in most parts of Africa, manifesting itself in the lack of purchasing power, political power, ill health, high child death ratio, low education, economic dislocation, personal violence, political activism, and natural disasters. Economic growth demands space, energy, and resources and puts stress on the global climate, waters, biodiversity, and vegetation. The erosion of natural and biological resources goes hand-in-hand with diminishing cultural diversity. Many traditional societies break up and numerous customs, cultural expressions, and languages are becoming eroded.

Despite global efforts to come to worldwide trade liberalisation, international trade relations are still far from equal. Export subsidies and import levies and limitations are still being used by the major economic blocs. Most tropical countries do not have equal opportunities for access to the global economy. Production now often takes place in parts of the world where over decades favourable opportunities have been created by a combination of protective measures and support mechanisms (subsidies, infrastructures, technology development, and extension) (Haverkort, van't Hooft, and Hiemstra 2003a, 29).

In many parts of Africa poverty predominates and many young people migrate to urban areas or abroad in search of greener pastures. Privatisation of health services and liberalisation of agricultural input supply tend to make health services and agricultural input services beyond the reach of many rural people. The result of these processes is that those areas or sections of the population that want to maintain and to further develop the local economy and cultural identity find it increasingly difficult to achieve their goals.

Hence endogenous development offers a paradigm shift for Africa's development. This shift has a niche in endogenous development from a cosmovision perspective (Haverkort, van't Hooft, and Hiemstra 2003b, 137). Why do I say so?

Supporting endogenous development

It is instructive here to look at the empirical findings of work done in respect of endogenous development by two NGOs in Africa, namely the Centre for Cosmovision and Indigenous Knowledge (CECIK – Ghana), and the Association of Zimbabwean Traditional Environmental Conservationists (AZTREC). These were implemented under the auspices of the Project Comparing and Sustaining Agricultural Systems (COMPAS) of the Netherlands. They include the following support activities that can be used to enhance endogenous development:

- Capacity development for identifying local resources and ways to access them;
- Understanding, testing, and improving local practices;
- Building on locally felt needs and values of different groups;
- Maximising local control of development;
- Enhancing the local capacity for learning and experimenting;
- Identifying new development niches based on the characteristics of each local situation;
- Optimising local resources through selective use of external resources;
- Retention of the benefits in the local area;
- Exchange experiences between different localities and cultures;
- Further understanding of systems of knowing, learning, and experimenting;
- Training and capacity building for rural people, development staff, and researchers.

Creating an enabling environment for endogenous development

Within the COMPAS network it was found that in order to stimulate a structural poverty reduction an enabling environment for endogenous development has to be created. Local initiatives may not result in effective changes if the wider legal, economic, or policy environment is not conducive to the results aimed at. Hence the following policy influencing and advocacy issues need to be addressed:

- Measures can be taken to secure intellectual property rights.
- Training of future experts in the right attitude and relevant knowledge will contribute to adequate human capacity development.
- Policy supportive research and dialogues between different actors can stimulate the adjustment of policies in prices, marketing, investment in education, research and development, legislation, etc. These can apply to sub-national, national, and international policies.

- Regional planning can lead to specific activities such as integrated nutrient management at regional level to link flows of nutrients, food, and biomass to other regional programmes.
- Networking and regional and national lobbying would allow the creation of an enabling environment for endogenous development above the farm or community level.

Through documenting and publishing experiences, intercultural dialogues can be held that allow for joint learning and co-evolution of different learning paradigms.

Code of conduct for enhancing endogenous development

The experiences of CECIK and AZTREC have shown that work with indigenous practices and knowledge as an 'outsider' implies certain risks. These include disturbing the status quo at community level, extracting local knowledge for purposes not in the interest of rural people, domination of local processes by outsiders, prying too much into people's private matters, and the introduction of lifestyles that are not consistent with local values. Therefore, in their work with the rural people CECIK and AZTREC have established the following code of conduct (Gonese 1999a, 1999b; Millar 1999a, 1999b):

- Accept the idea that local communities have indigenous knowledge systems with their own rationale and logic, and be prepared to learn from them.
- Commitment to work in the interest of the local communities. Programmes will only be implemented after approval of the local community and its leaders based largely on the norms and values of the peoples themselves – bio-cultural protocols.
- Accept the rules and regulations set by the local community for attending and receiving visitors, and respect the limitations set by local leaders.
- Accept and seek complementarity between external knowledge and local knowledge systems. Avoid the domination of external over local knowledge and value systems
- Accept that in many cases new methods will have to be developed, as the conventional approaches for research and development may not be the most appropriate.
- Project staff should be predisposed to (re)learning when they engage with local communities. Learn empathetically from the local knowledge system, analyse it, and enter into a respectful and constructive dialogue about the positive and negative aspects, the possibilities for improvement, as well as the epistemologies and paradigms.
- Accept the guidance of local leaders to ensure that the information collected will be in the interests of the community, thus respecting traditional intellectual property rights.
- Accept the importance of exchange of experiences within and between rural communities. Publish experiences for other audiences only after approval of the communities involved.

Working within endogenous development

This section primarily tries to respond to the needs and questions asked by my students in the University for Development Studies. The students legitimately asked for empirical work done in Ghana or Africa in respect of endogenous development and the new challenges posed by this paradigm. My own previous and current work offers useful practical resources and learning tools for my students and other development practitioners.

In my earlier works (Millar 1994, 1996, 2003, 159) I established that **empathy** is the guiding principle for re-orientation of the "self" in order to build a trustworthy relationship with rural people and in so doing, engage in constructive dialogue. Empathy means putting oneself in the position of the others and attempting to understand and appreciate how the "other" thinks and feels. Within this context CECIK has evolved steps to conduct learning from the point of view of empathy, which is strategic in conducting endogenous development. The steps include the following:

Building empathetic relationship with rural people

- Identify the population concerned: what is the boundary of the target population, their tribal name, socio-economic description, and location?
- Compile existing anthropological, agricultural, and other relevant literature and summaries of existing information on the above, consulting the libraries in the area in the first instance. Possibly libraries in other countries must also be used. In many cases, for example, anthropological data are more easily retrieved in the West than in the country they refer to.
- Cultural expressions in houses, painting, sculpturing, and religion can give important information.
- When entering into a new community, make sure to show respect and interest in their values. When already working in a community, you should consider efforts to redefine the existing relationships.
- Agree with the population on the activities. Agree on the goals and activities to be carried out as well as on the roles of the different leaders. As much as possible make a covenant with the community on the process and ownership of the results. Use the principle of prior informed consent and joint planning.
- Agree about the methods to be used for learning about indigenous knowledge and indigenous institutions. Consider (a mix of) the following methods: asking key persons about the traditional social structure and leadership; interview traditional leaders (men and women), be keen to learn about their worldview and about their role in teaching and experimentation; take note of oral life histories; use village workshops, village theatre, visual presentations, and linguistic analysis; understand folk stories, creation myths, songs, customs, rituals, visual expressions in painting, architecture, and sculpture; participate in festivals, rituals, and other important events. Accept that you are a learner.
- Persons to be interviewed can be traditional leaders, spiritual leaders, healers, old farmers, young farmers, or key persons with specific information.
- In all cases make sure that gender differentiation is being made.
- Make sure that the results of this process are discussed and assessed with members of the community and that conclusions drawn together, taking into account gender, age, and power differences in the community.
- Ensure, as much as possible, that documentation is in a form that is in line with rural traditions and can be retrieved and used by the population.

Framework for experimenting within endogenous development

Experimenting or testing is a very important component of endogenous development. A framework based on empirical findings of COMPAS partners, of which CECIK is a part, over a four-year period, is useful in this respect (Millar 1993, 2003, 153). To develop a framework for experimentation and testing, the following questions are suggested:

(1) How to (re)enter the self-development process of the community and build up a good working relationship?
(2) How to develop a working agenda together with the community?
(3) How to determine the parameters for testing and experimenting together?

(4) What common strategy can be used to realise the common objective? And how to design the experiments?

(5) How to keep track together of the process (monitoring, assessing, assisting, supporting, sharing, and developing)?

(6) How to judge or determine the results or products of the process, considering the three aspects of spiritual, material, and social growth?

(7) How to leave their process such that the community can continue the growth and self-development process (to continue and to up-scale, considering the spiritual, material, and social equilibrium of the community)?

The experiences of CECIK can be used to further deepen the understanding of the *how* to conduct experimentation.

1. *How to (re)enter the process of self-development of the community and build up a good working relationship?*
   - If a field worker is approached by somebody from the community with a request for help, it is important to find out what is it that the community wants and what can be considered good experiences so far. An appointment can be made during which the fieldworker should be open to existing knowledge and the institutional context. Openness and transparency about the interest in cosmovision including its spiritual aspects is also needed.
   - In case the process is renewing older contacts, make sure to follow-up on a previous dialogue. It is important to make clear to the community how cooperation is foreseen, also indicating the interest in cosmovision and its three dimensions (natural world, social world, and spiritual world).
   - When possible and relevant, participation in ceremonies and festivals is desirable, thus learning about the cosmovision, indigenous institutions, and cultural identity of the community.
   - Familiarisation with the cosmovision and the culture of the area by studying previous cosmovision, anthropological, and development studies can provide important background information.
   - It is important to establish contact and have discussions with traditional and spiritual leaders, spirit mediums, healers, and elders.
   - When meetings are organised, reflect also on traditional ceremonies and rituals as performed by traditional and spiritual leaders.
2. *How to develop a common agenda?*
   - Starting with positive experiences on the issues brought forward (yours and theirs).
   - Consulting spiritual institutions.
   - Meditation and reflection.
   - Using participatory diagnostic tools.
   - Carrying out visits and cross-visits.
   - Having extensive dialogue with the community.
   - Using local means of communication such as drumming, music, drama, design on houses, textiles, and art objects.
   - Participating in ceremonies as a way of engaging with interest groups/stakeholders.
   - Being aware of possible resistance, conflicts, and confrontation and respect limits set by the community.

3. *How to jointly determine parameters?*
   - Ask the spirit mediums about their parameters and to what extent is it possible for them to carry out and be involved in experiments.
   - Identify the successful parameters in the community's experiences.
   - Evaluate project parameters against those of the indigenous cosmovision.
   - Use a timeframe that respects ritual calendars and astrological data.
   - Take into account the location: respect sacred places and other qualities of the location indicated by their cosmovision.
   - Take into account a resource frame that responds to local perceptions and value system.
   - Incorporate socio-cultural issues such as taboos, totems, class, and caste, and involve local authorities in decision-making and management issues.
   - Be gender sensitive and make gender and generational differentiation in experimental parameters and indicators.
   - Incorporate spiritual elements as indicated by the spiritual leaders:
   - Respecting signs of the ancestors, indications by dreams and visions.
   - Deal both with qualitative and quantitative data.
   - Gradually develop gradually criteria, indicators, and strategies.

4. *How to design actions?*
   - Use traditional institutions in the design process and if suggested include indications given by dreams, visions, and intuition.
   - Being open for modifications of the design during execution.
   - Let the local institutions be accountable for the design (for the whole action).
   - Start from the known to the unknown.
   - Evoke indigenous ways of designing and experimenting.

5. *How to monitor?*
   - Build the strength to monitor in the communities and traditional institutions.
   - Use relevant monitoring components from tools of participatory approaches.
   - Let the spirit mediums be part of the monitoring team.
   - Set up community codes of conduct and bylaws.
   - Also make observations and reflections by meditation.
   - Conduct participatory dialogues and use opportunities offered by ceremonies, festivals, and other occasions as well.
   - Use field notebooks and diaries, community registers, indigenous recording forms, village albums, reports, and audio-visuals.

6. *How to judge the results of the experiments?*
   - Revisiting the parameters with all stakeholders.
   - Make a comparison with project baselines and the community's baselines established at the beginning.
   - Be open for intended and unexpected results and modifications of the experiments by the community.
   - Have open discussions on the results: is everybody satisfied? Can results be felt, tasted, parameters measured? Can symbols or signs indicate the results?
   - Spiritual results can also express themselves in the physical world.

7. *How to leave the community?*
   - A periodical follow-up with decreasing intensity, engage in a social conversation on the subject of phasing-out.
   - Consulting spirit mediums.
   - Being sure that the process is embedded in the culture, especially when you replicate or scale-up.

- Still being part of the community, especially for festivals and ceremonies.
- Ensuring networking with traditional institutional networking throughout.
- Developing reciprocal arrangements with similar communities.
- Involving women and the youth early, looking for signals from them and asking for their commitment.
- Having various forms of documentation available locally. Leave summary behind as a point of discussion.

References

Gonese, C. 1999a. "Three Worlds." *Compas Newsletter* 1, February: 24.

Gonese, C. 1999b. "Culture and Cosmovision of Traditional Institutions in Zimbabwe." In *Food for Thought*, edited by B. Haverkort and W. Hiemstra. London: Zed Books.

Haverkort, B., and W. Hiemstra. 1999. "COMPAS: Supporting Endogenous Development." In *Food for Thought*, edited by B. Haverkort and W. Hiemstra. London: Zed Books.

Haverkort, B., K. van't Hooft, and W. Hiemstra. 2003a. "Cultures, Knowledges, and Developments, a Historical Perspective." In *Ancient Roots, New Shoots: Endogenous Development in Practice*, edited by B. Haverkort, K. van't Hooft, and W. Hiemstra, 11–28. London: Zed Books.

Haverkort, B., K. van't Hooft, and W. Hiemstra. 2003b. "The Compas Approach to Support Endogenous Development." In *Ancient Roots, New Shoots: Endogenous Development in Practice*, edited by B. Haverkort, K. van't Hooft, and W. Hiemstra, 29–36. London: Zed Books.

Haverkort, B., D. Millar, and C. Gonese. 2003. "Knowledge and Belief Systems in Sub-Saharan Africa." In *Ancient Roots, New Shoots: Endogenous Development in Practice*, edited by B. Haverkort, K. van't Hooft, and W. Hiemstra, 137–180. London: Zed Books.

Millar, D. 1993. "Farmer Experimentation and the Cosmovision Paradigm." In *Cultivating Knowledge: Genetic Diversity, Farmer Experimentation, and Crop Research*, edited by A. W. de Boef, et al. London: Intermediary Technology Publications.

Millar, D. 1994. "Experimenting Farmers in Northern Ghana." In *Beyond Farmer First*, edited by I. Scoones and J. Thompson, 160–164. London: IIED.

Millar, D. 1996. "Foot Prints in the Mud: Re-constructing the Diversities in Rural People's Learning Processes." PhD thesis., Wageningen Agricultural University, The Netherlands. Wageningen: Grafisch Service Centrum.

Millar, D. 1999a. "Traditional African Worldviews from a Cosmovision Perspective." In *Food for Thought*, edited by B. Haverkort and W. Hiemstra. London: Zed Books.

Millar, D. 1999b. "When the Spirits Speak." In *Food for Thought*, edited by B. Haverkort and W. Hiemstra. London: Zed Books.

Millar, D. 2003. "Improving Farming with Ancestral Support." In *Ancient Roots, New Shoots: Endogenous Development in Practice*, edited by B. Haverkort, K. van't Hooft, and W. Hiemstra, 153–168. London: Zed Books.

Van der Ploeg, J. D., and V. Saccomandi. 1993. "Impact of Endogenous Development on Agriculture." Unpublished paper.

Van der Ploeg, J. D., and G. van Dijk. 1995. *Beyond Modernisation: The Impact of Endogenous Rural Development*. Assen, The Netherlands: Van Gorcum.

African family values in a globalised world: the speed and intensity of change in post-colonial Africa

Charles Banda

Guest editors' note

The reflection that follows is unusual in a peer-reviewed journal, but it is here because the shaping of development policy and practices comes not just from data analysis, but also from the stories of people who are experiencing the psycho-social and cultural traumas many face as they are thrust into a globalised world. We think of the colonial experience in Africa, particularly after 1885, as having devalued and destroyed African history and culture. This story from Charles Banda, born just as independence was coming to Africa, suggests that whatever damage the colonial powers did to African societies, the process of destruction may have, in fact, accelerated in post-colonial Africa. The overlay of new structures, institutions, and cultural practices from the colonial period – for example, the adoption of Christianity – did not totally displace African traditions. African culture may have to some degree been insulated by indirect rule and other colonial governance patterns. Post-colonial Africa has experienced turbo-charged change, as multiple effects of globalisation, including information technology, migration, and commercialisation, crossed national borders and eroded local ways of doing things. Banda, speaking as one who has experienced this, reflects on what this rapid change means for African families and communities. This is a personal story, not often heard as we consider "development". We could find many other such personal reflections, different in the details, but similar in the concerns of disorientation from one's heritage. Though the contexts may differ, the concerns may not be so different from those of certain generations in the developed world, who are also seeking to find meaning in a globalised world.

Still, this does not answer the question of why this reflection should appear in a journal of development practice. As older generations mourn the loss of traditions and practices that regulated behaviour and gave life meaning, or the loss of language, religious traditions, and communal identity, new generations are thrust into an individualistic market society. Banda's reflections offer clues on how to draw on the past to build the future. The postscript addresses this.

Background

My name is Charles Banda. I was born in 1958 in the then Northern Rhodesia. Northern Rhodesia was a British Protectorate until 1964 when Zambia became independent. I am the second born in a family of 10 children. The family was composed of six boys and four girls. Four of my siblings plus my father have passed on.

Context I grew up in

My memories of growing up are generally fond ones. My recollection of my childhood begins from the time I started school in 1964, just when Northern Rhodesia was attaining its independence from Britain to become Zambia. I was born into a family of Malawian migrants who had gone to Zambia to seek employment. My father had attained standard six level of education in South Africa (where his parents – my grandparents – worked in the Durban sugar estates).

In 1954, the year that he began to work in the Roan Mines, in the Copper Belt, there were very few Africans who had his level of education (standard six). He not only spoke fluent English, but also good Afrikaans because of his South African schooling. Most of the senior managers running the mines were white South Africans and because of this, I believe my father found favour as he could speak their languages.

On attaining independence, Zambia had opened up what were once exclusive white schools to all races and my sister and I were among the few Africans to be enrolled at one such school, called Harrison Primary School.

Despite having spent considerable time at school in South Africa, the memories that I have of my father are that he was as traditional and conservative as anyone who had remained in the village back in Malawi. Although we lived among Zambians and shared everything in the way of life, my father always reminded us that we were Malawians. My father was a disciplinarian who raised us in Christian values.

Extended verses nuclear family values

Growing up in our family, we quickly realised that being our dad's children did not make us more special than our relatives who came from Malawi to visit us from time to time. Our mother was an ordinary housewife, whose role was to take care of us at home and ensured that we were fed. It was our father who shaped our character and who we are today.

My father instilled in us a sense of belonging, that blood was thicker than water. What he meant did not just refer to us as the nuclear family, but that all relatives were as important as any one of us and for that reason they were treated just like any of us. No priority was visibly given to us above the extended family. My father believed in an approach of "if you scratch my back, I will scratch yours". He would do everything possible to make sure that relatives felt as at home in our house as we children did. He believed that all relatives were important and therefore were to be treated as special. He believed that a good turn deserved another, that if you were good to others, they would be good to you. He was also strongly convinced that relatives, especially extended family relatives, had a role to play in instilling moral values. Our father's devotion to the extended family never ceased to amaze us. Each time my father received a letter from Malawi telling him that his elderly mother (our grandmother) was sick, he would immediately take leave from work and jump on the next bus to Malawi to see her. It would be upon his return that he would explain or offer a kind of apology for having gone without proper preparations. Over time, we had come to realise that the subject of our grandmother and the place she held in his heart was not discussable. Today when I reflect on this seemingly odd behaviour, what I see is a deep love and connection between mother and child. I do not think that we parents of today would love our parents in the ways our fathers and mothers did in the past; nor would our own children love us in the same way as my father did his mother. My father would sacrifice the last penny in the house to pay for transport to see his mother. Of course, my father was brought up by a single mother, having lost his father as a teenager.

In the past, the naming of children was an extended family responsibility, illustrating the value placed on extended family, and the birth of a child was celebrated by the whole clan or community. For this reason, names of children came from elders within the family and not necessarily from the parents of the child. In the past, children born in a family or clan were seen as replacements for those that had passed on; so new arrivals were often given the names of those who had deceased. This symbolised continuity and the dead were regarded as living in the new arrivals. It was not surprising in those days to see young ones taking on the character of the deceased person for whom they were named. One would often hear people saying, "he is a good hunter like the late so and so". Today, the naming of children is done by the biological parents, and the names given do not have any symbolic value in the communities or extended families from which the children come. Today children may be named after celebrities such as Jay Z or Beyoncé. Obviously, these are a result of the global mass media influence.

Key family values absorbed from my parents

Despite the fact that we grew up in a town, our parents taught us great values that I personally cherish and would love to share with my own children. Our parents taught us that, although they were our biological parents, we still had to treat any elder person with the same respect that we accorded our parents. Morally, our parents imparted a number of values aimed at our proper upbringing. These values ranged from general behaviour to sexual conduct. Different methods were used for boys and girls in imparting values. During puberty for boys, usually around 14 years upwards, all boys in the extended family and even the village were expected to stay in a specially erected boy's communal home. Boys were expected to stay in this home until they acquired their own homes after marriage. These homes served as a place for education on life values. It is in these homes that boys would be taught life-long skills, including being introduced to sexuality and preparation for adulthood and parenting. Throughout my teenage school holidays, I would be brought to these homes in Malawi thousands of miles away from our home in the Copper Belt in Zambia. On one of the nights in the teenage home (Mphala), we were privileged to have a respected uncle come in and talk about hygiene. He talked about many issues regarding hygiene but when he came to the issue of cleanliness, he mentioned that it is always important to keep oneself clean.

Although I personally did not have much experience of these homes as I only came in once in a while, I was told that these home provided the much-needed life experience that is missing today in the ways that adolescent boys are raised. It was in these homes that boys were prepared to become men. Boys were taught different roles, including those of being a husband (taking responsibility, and respecting women).

Like boys, girls on attaining puberty were often taken away from their parents' homes to live with their aunts. At the first sign of menstruation, girls would go through a process of learning how to become a woman, wife, and mother. During this period, the girls would be taught how to handle themselves whilst going through the menstrual period, how to treat their husbands, and what to do when they became pregnant. I am told that girls were sometimes given lessons by elders within the extended family about appropriate behaviour in marriage and sexuality.

As children growing up, my father was strict that always we did school homework every day between 4pm and 6pm, and we showed him our school work together with the homework done. He normally would check our day's work soon after supper (as he did not want to hurt anyone before they ate). He believed that any interventions or punishments should be administered after children had eaten. If this was done before, it was likely that children would not eat. This gave us a spirit of hard work and at the same time it taught us that he cared about our schooling. Sometimes, my father would mark our work and somehow this did not offend our school teachers.

Our parents taught us that we had to be respectful to all adults. This was to be shown through, for example, kneeling when offering something to an adult. Young boys under the age of 14 are expected to kneel like girls. Adult boys were not expected to kneel but squat or to show respect by holding the cup or whatever they were presenting with two hands rather than one. Girls were expected to kneel down regardless of age, to not play near adults, to not shout in the presence of adults, and to eat last when eating with adults. My mother took the responsibility of teaching the girls to become women, while my father took the responsibility of building men out of boys.

We were taught that furniture in our home belonged to the family but as children we were not allowed, for example, to play with the radio lest we damaged it. My father and mother had bicycles, which we rode after asking permission. We were not allowed to play with things (that were expensive) without seeking permission. It was considered disrespectful to take anything without asking for permission. We could not, for example, eat bread left over from morning breakfast without asking for permission first. My parents never said no if we respectfully asked for something. A "no" answer only came if they felt that we would hurt ourselves or damage the property in question.

As we grew, we were told that boys could not play with girls. This marked of the beginning of the sexual orientation training. We were told that if a boy played with girls, he would develop breasts. I had seen a number of males apparently who physiologically looked like they had breasts, and I did not want that so I kept away from girls. The subject of sexual orientation was a tricky one. It is one area that I feel my parents felt shy to dwell on it and left it to other extended family members to deal with.

When I was about 14 years, my uncle told me that girls "*can burn you*". I did not want to be burnt. I kept away from girls until college when the whole thing was demystified. In the past, when we were growing up, using myths for educating the young was common. The importance of such myths, in hindsight, is that they enabled youths to concentrate on other important life issues rather than those that the elders felt could derail the youth from attaining their full grown-up potential. Sex was for adults, and to ensure that this remained so, the burning myth was used. During our time of growing up, there were very few teenage pregnancies. People became pregnant in marriage. Marriages themselves were arranged by elders in the family. Once adults felt that one was ready to marry, they would arrange for a woman, who also was ready to get married, from another village. For one to be ready to marry, you had to prove that you were a man. This was done by proving that you could maintain a family, which implied that you had to have a garden from which you could harvest your maize, cassava, or millet, etc. A hard working boy was likely to marry earlier than a lazy one. Girls who were respectful and hard working in the village, too, were also more likely to find a husband than those who were not. I should state here that in the past, people tended to marry when they were older than today. In the past, most people getting married were in their late 20s or early 30s. The positive thing with this as opposed to today is that people were entering marriage when they were fully matured. Such people did not have very large families because they were already much older. Today, girls as young as 13 years old have children. This may explain in part the population explosion in the world today.

It should be said again that a hard working boy or girl's reputation would travel far and wide. They were never short of a husband or wife. As for me, my parents arranged my marriage with my wife, Alice, who was also born in Zambia like me, and moved with me to Malawi. We are very happily married and have four children.

One of the things that I still admire about my family and about growing up in the small town of Luanshya is the solidarity in the family. Although all of us were born in Zambia, thousands of kilometres away from Malawi, we all spoke our ancestral traditional Malawian Tonga language. Although we also spoke the Zambian Bemba language, we were not allowed to speak this in our home. As a result we were able to retain a Malawian identity despite having been born outside Malawi. This I think is the reason why, when my father retired in the early 1980s, all of us returned with him to Malawi.

Values taken on to my family and challenges experienced

I am married to my lovely wife Alice, whom I met through my sisters. We got married in 1982. Like me, Alice was born in Zambia from Malawian parents. My parents knew Alice's parents. My parents were very impressed with Alice's behaviour, i.e., that she had completed secondary school, she used to help her parents on home chores, and she was well behaved. These character-istics led my parents to arrange a meeting between the two of us. It was love at first sight. Although I left Zambia and Alice behind to return to Malawi in 1981, Alice followed me to Malawi the following year and we got married in church in 1982. The following year we had our first baby boy. We have four children, one boy and three girls.

This kind of marriage is a typical, arranged marriage. I am happy that I am married to Alice. Our marriage is very strong because it is a bond between two families, the family of Alice – the Phiris – and my family, the Bandas. Our family is strong as it has the blessings and support of the two families. We have a very strong support system, composed of the two families and their extended family members. For the two of us, this has been a great source of inspiration as we have been and are still seen as role models for the two families. For this reason, we would not like to let the two families down. This does not in any way mean that we would hide serious dirty linen under the carpet. Any disagreements or misunderstanding that might occur, if any, would be handled by us or the two families. So far, having been married for almost 35 years, we have not yet had any reason to call upon the extended family to mediate in our problems.

Today the world is different. Children will be very reluctant to get married to someone they do not know very well, let alone enter into an arranged relationship. This, of course, is good. I cer-tainly do not agree with arranged marriages, where one party is under age and is in no position to make decisions of his or her own. Marriages today are strictly the affair of two people. In my mar-riage, my parents and my wife's parents are key stakeholders in our marriage, which is not the case with our daughters. Married couples today would rather divorce than seek mediation or share their marital problems with their parents or extend families. This is because some of these married couples went ahead and got married without the consent of their parents or extended families.

Today the world is so different. Thirty years ago, we did not have the Internet or social media. Today a child of 10 may know much more than an adult of 20 or 30 years ago. Most of our chil-dren no longer rely on us for the wisdom or values that will guide or mould their character.

Despite this, there a number of values that I borrowed and cherish from my own upbringing that I have either dropped or modified and taken into my family. Issues of respecting adults, not being near adults unless called, showing interest in the school work of the children, etc. I greatly appreciate the fact that if it were not for my father's interest in my schooling, my siblings and I would not be where we are today. Right from the time that my father saw an opportunity to get us into a good school, he made that arrangement without hesitation, despite the fact that he may not have had adequate money to pay the fees. He sacrificed a lot in order to ensure that we attained a good education.

Today, my sexual discussion with my daughters and son is limited because I know they may know much more than I do because of the Internet and social media. My role is to instil in them the need for getting an education and then marriage afterwards. This is what I have dwelt on. In Malawi and most other African countries, girls tend to get married early. For this reason, very few girls have managed to get a college education and live a respectful life independently.

The role of my mother was basically to raise and take care of us as a family. My mother was a typical family woman who never worked. Her work was simply to take care of us children and our father. Despite her calmness, my mother had salient powers, which she called upon when need be.

Whilst my father was a strict disciplinarian and to some extent an autocrat, our mother was always opening up avenues of alternatives. We, for example, were not allowed to switch the radio or TV on if our father was not around. My mother would allow us to use these gadgets in the absence of our father. My father gave us a *flock* when we did wrong. Our mother would determine how far the beating could go. If the punishment was too much, she would order our dad to stop and he would do just that without question.

As children, we were raised in an extended family set up, not only by our biological parents but by the whole clan or extended family. This meant that our relatives, especially the elders, had an input equal to our parents in our upbringing. When I was a teenager, I was often sent to visit my uncle. During my stay with my uncle, I was oriented to my expected roles as a parent. This is something that is not now happening. My son has just married a woman of his choice. We love them both. I was never consulted regarding their marriage. We as parents were simply informed by our son that he was in love and was going to get married. The couple have a beautiful daughter. We love them very much. I guess there is nothing much that we could have done or envisage to do to except to love them unconditionally.

Are the values going to be taken forward? How satisfied am I with this?

I am certainly positive that some of the values imparted will be taken forward and passed on to the next generation. This is so because my daughters and son have focused on creating a future for themselves rather than seeing getting into marriage as the most important thing. My daughters, all who are now above 20 years old, are respectful to all adults and I feel they now understand why I did and said certain things. Today my relationship with my children is that of an adult to another adult. We can talk about any challenges they face as adults. This is something that I never did with my own parents.

Family life today compared with the past

Today is certainly different from the past in which I grew up. Whilst I endured a beating from my parents for mistakes done and was able to appreciate why they did what they did, today, I cannot administer corporal punishment because my child has failed to do his or her homework. Today, the world is a global world; in the past my parents treated us as children from the village and wanted us to have the same traditional values that children in the village had. It was important for parents to bring up their children in their tradition. These things no longer exist, the world is a global family and Western ideologies of rights and responsibilities now reign. By law you cannot punish your children. Whilst this is the norm and seen as good by many, I feel a little spanking creates fear in the child to do right next time. Dr Hastings Kamuza Banda, the first pre-sident of Malawi, always advocated that children and the youths must be guided at all costs. Though he did not say spank them, I believe it is implied. Under his leadership, Malawians were hard working and disciplined people. I am not sure the same can be said of them now.

Another challenge that I observe today is that most children are born of parents who were born in towns, where traditional values have been greatly eroded or abandoned. Most of these families are nuclear families and have little or no contact with villages or communities where their fore-fathers came from. Such families have adopted a Western type of life. Their values are basically copied from the mass media. The chief culprits are television and print media, which have por-trayed African values as primitive and outdated to some extent. Because of this bombardment, the new generation of African parents are abandoning their traditional ways of life and adopting the Western type of life and values. Traditional African values as I knew them are fast vanishing in most African countries, including Malawi, because most new parents were born in town and never experienced traditional values. These town parents and their siblings know nothing else but the "global values" as preached by the mass media.

Most traditional values have become extinct, except for a few such as celebrating the wedding of children. Here again, parents of both the girl and boy will come together to celebrate this union. The birth of a child is another that has remained an important occasion since it symbolises an arrival to and perpetuation of the family tree. Regardless of how the family started, both families of the couple will celebrate the arrival of new ones in the family. Death is also another important occasion where all family members come together to share the burden of grief. I cannot imagine what would happen if death was left to one family alone to handle. These rituals or ceremonies help to bond families and build relationships. In the African situation, death is regarded as an important occasion where all family members must come together and bury the dead relatives. It does not matter how far away you are, the family members will wait for you before the burial can be made. Every kinsman must be given an opportunity to say farewell to the deceased. Every relative of the deceased is given the opportunity to see the face of the deceased for the last time before burial.

Key challenges facing African families and parenting

The most visible and profound challenge today facing the African family and parenting is the breakdown of the value of the extended family. In the extended family, children were raised by the whole community. Children belonged to the community. Today this no longer exists. What this means is that each parent is responsible for ensuring the growth and well-being of their children. In the extended family children are taught different values of life, and are basically guided on what is right or wrong in life through support from uncles, aunts, nieces, nephews, cousins, etc. In the nuclear family, this kind of support is lost. The children may not even know their next of kin beyond their blood brothers and sisters. Educating children, both girls and boys, in a nuclear family is left to their biological parents who may not be good at it. It is a big challenge for biological parents to talk to teenage girls about sexual health issues, for example. Because of the absence of this extended family support system, children tend to find support elsewhere, and usually through peer influence or the mass media. In the past, it was unheard of to hear that a girl as young as 13 years old was pregnant. Today, this is widely happening.

My advice for success of African family life and parenting

There is need to carry out research into family life and parenting as it used to be in the past. The challenge with Africa is that very little has been documented in terms of past family life. Very little is known about the role that the extended family played in the lives of the people in the communities. From the research results, we can decide what may help us in becoming better people today and in the future

At the rate that Africa is losing its rich traditional values, I wonder how this can be preserved. This could be one of the roles that NGOs in development can work on – preserving the positive values of Africa that have been passed on from generation to generation. The issues of HIV/AIDS might not have reached the proportions we are seeing today if some of the traditional cultural behaviours of the past had been imparted to the new generation. Churches and mosques can also play a key role in passing on traditional cultural values that promote positive behaviour.

Guest editors' postscript

A close listening to what Charles Banda has to say can yield an understanding about how to ground community development and social programmes on traditional values. It is important to listen to the spirit, not the specific examples of traditions that Charles Banda mentions. When he talks about puberty-age boys going to a community home to be taught about hygiene, sexuality, adulthood, and parenting, or puberty-age girls being taken to an aunt to learn about appropriate behaviour in sexuality and marriage, he was not calling for exact replication of these traditions. He suggests drawing on the spirit of the tradition, which is that the community has a responsibility for providing knowledge about sexuality and setting and enforcing behaviour standards for young people as they go through a difficult transition to adulthood. In a world of HIV/AIDS prevalence, commercial messages glamorising sex, and an older age of marriage, community influence may be far more powerful than that of parents alone or of distant government services in protecting and preparing young people for adulthood. Myths, such as "girls can burn you", can be replaced by knowledge of the consequences of HIV/AIDS or early pregnancy. Equally when Banda talks about the different roles of boys and men versus those of girls and women, he is not calling for a subjugation of women's rights, but for roles of women and men that are respected and honoured, and for family members who support each other. Charles sees the world today as *"certainly different from the past in which I grew up"*. He argues that we need to study Africa's *"rich traditional values"* so that *"we can decide what may help us in becoming better people today and in the future"*.

African philanthropy, pan-Africanism, and Africa's development

Bhekinkosi Moyo and Katiana Ramsamy

Reflective and theoretical, this article explores the foundations and principles of African philanthropy and juxtaposes them with pan-African-led development. It pays particular attention to new continental initiatives, such as Agenda 2063. It points out that African philanthropy, by its definition and practice, is the foundation for development. This is because the identity of an African is premised on philanthropic notions of solidarity, interconnectedness, interdependencies, reciprocity, mutuality, and a continuum of relationships. No one embodies these better than Nelson Mandela in his demonstration of the link that exists between pan-Africanism and African philanthropy in the development process.

Cet article, qui adopte une démarche réflexive et théorique, examine les fondations et les principes de la philanthropie africaine et les juxtapose avec le développement impulsé par toute l'Afrique. Il accorde une attention particulière aux nouvelles initiatives continentales, comme l'Agenda 2063. Il fait remarquer que la philanthropie africaine, de par sa définition et sa pratique, est la fondation du développement, ce parce que l'identité d'un Africain est fondée sur des notions philanthropiques de solidarité, d'interconnexion, d'interdépendances, de réciprocité, de mutualité et un continuum de relations. Personne n'incarne mieux ces notions que Nelson Mandela, dans sa démonstration du lien qui existe entre le panafricanisme et la philanthropie africaine dans le processus de développement.

El presente artículo, reflexivo y teórico, examina los fundamentos y los principios de la filantropía africana, yuxtaponiéndolos con el desarrollo impulsado a partir del panafricanismo. En este sentido, destacan en particular las nuevas iniciativas promovidas a nivel del continente, como la Agenda 2063. Asimismo, el artículo señala que, tanto en su definición como en su práctica, la filantropía africana constituye la base del desarrollo, pues la identidad africana responde a ideas filantrópicas de solidaridad, interconectividad, interdependencia, reciprocidad, mutualidad, así como de un continuo de relaciones. Nadie encarna mejor estas ideas que Nelson Mandela, quien durante su vida resaltó la vinculación existente entre el panafricanismo y la filantropía africana a nivel del proceso de desarrollo.

Introducing African philanthropy in development

"During my lifetime I have dedicated myself to this struggle of the African people. I have fought against white domination, and I have fought against black domination. I have cherished the ideal of democratic and free society in which all persons live together in harmony and with equal opportunities. It is an ideal which I hope to live for and to achieve. But if needs be, it is an ideal for which I am prepared to die." (Nelson Mandela, 20 April, 1964)

These words by Mandela, and his actions throughout his life, aptly capture the possibilities presented by African philanthropy and pan-Africanism. Throughout his lifetime, Mandela succeeded in achieving most of the developmental quests that any society aspires for: liberation, freedom, democracy, and social justice. In 2008, at the 6th Annual Nelson Mandela Lecture, he said, "*a fundamental concern for others in our individual and community lives would go a long way in making the world the better place we so passionately dreamt of*" (12 July, 2008). He demonstrated this concern for others, and what stands out across his life is his service to humanity and selfless sharing of his life with everyone, including giving of his money, time, and kindness, and even establishing philanthropic institutions to further these ideas. After his death in 2013, his will (which was read publicly) had all the marks of philanthropy. Not only did he give his estate to many deserving causes, he also demonstrated that it is possible for African philanthropy to be the foundation and anchor for development. We argue here that Africa must follow this example if indeed it wishes to progress and achieve sustainable development. For this reason, new African initiatives that build on African philanthropy or its underlying values – such as the Agenda 2063, the African Union Foundation, the African Grant Makers Network and others – should take a page from Mandela's book on how he lived his life and brought about the developmental imperatives he sought.

Even though there is a gap between the ideal and practice, we are of the view that it is time to revisit the link between pan-Africanism and African philanthropy, given the urgent need for solidarity in resolving conflicts and the opportunities for development in the current economic, social, cultural, and political trends in Africa. Given the current disregard for human dignity in most parts of Africa, especially by the political elite, and the recorded failure of past pan-African projects, we are aware that not everyone will agree with our optimism. Yet we are clear that the pan-Africanist vision is not lost yet; in fact, it is seeing a revival. We are also not arguing for a type of pan-Africanism that ignores the various fundamental differences in history, culture, societal systems, and values across the continent; rather, we recognise diversity and argue for its utilisation as a unifying force for development in Africa. Our intention here is to outline a vision for the implementation of development initiatives from a pan-African and African philanthropy perspective. Additional research will be needed later to establish the challenges and barriers to such an approach, but for the purposes of this article we provide a theoretical reflection and conceptual framework of an African-led development, anchored in principles and values enshrined in both pan-Africanism and African philanthropy. We argue that development cannot be achieved outside the parameters of these values, for through their accountability, transparency, and empowerment components they address one of the vexing issues in African development: that of governance. In this context, we propose that the various African initiatives mentioned must position themselves as beginning points and frameworks for development in Africa. There will be challenges, such as lack of political will, inadequate resources, competing agendas, and incapacities; but these need not be an excuse for not getting on with the agenda of developing Africa.

This article therefore seeks to position African *philanthropy* – a term that was once foreign in Africa, even though its practice has always been a reality, and one that many of us have grappled with for many years to make contextually relevant – at the heart of Africa's development trajectory. Development ought to be transformative, sustainable, and essentially based on Africa's own institutions, informed by its own knowledge systems, and supported by its resources. Because the lead author has written extensively on African philanthropy and its practice (including definitions and the status of African philanthropy, as in Moyo 2009a, 2009b, 2010, 2013, Moyo and Aina, 2013 and others),[1] this article does not duplicate these pieces. Rather, the focus is on the core argument relating the key confluences of African philanthropy and pan-Africanism, to position philanthropy as a paradigm for development.

Some brief distinctions in terminology are useful: generally, *philanthropy in Africa* refers to forms of philanthropy that may not necessarily be African but are operational in Africa. Examples include international forms of philanthropy such as foundations that are present in Africa but with origins outside of Africa. *Philanthropy with African features* is any philanthropy that is not bound or limited to a geographic space called Africa but is essentially structured in a value-system that resembles that of the African format and nature. *African philanthropy* differs from these, and is outlined briefly here.

> "Though not a common or even user-friendly concept in Africa, philanthropy as a phenomenon perhaps is best captured by the notions of 'solidarity and reciprocity' among Africans and some of the features that accompany relational building. As a result, therefore, culture and relational building are central attributes in defining what philanthropy in the African context looks like. Philanthropy is intrinsically embedded in the lifecycle of birth, life and death of many, if not all Africans. At any one given time, one is either a philanthropist or a recipient of one kind or another of benevolence." (Moyo 2011, 1)

Historically, philanthropy as understood to mean "love for humanity" has always been practised by Africans in their different and unique contexts:

> "Understood mainly as giving or helping (in the narrowest sense), or even better more encapsulated as solidarity and reciprocity – this entailed collective or individual efforts towards a social or public good. This conception of good was not divorced from questions of well living, welfare or wellbeing – understood today more in terms of sustainable and people driven and inclusive development." (Moyo 2011, 1)

It is important to debunk the notion that African philanthropy is informal, indigenous, and "traditional". Such an analytical or conceptual framework can have the negative effect of perpetually condemning African initiatives and frameworks to the margins. If this happened, the very essence of an African and its development initiatives would be undermined. Of African philanthropy, Moyo writes:

> "Due to analytical influence and frameworks primarily from the West, philanthropy in Africa or, to be more specific, African philanthropy, has sometimes been wrongly and maliciously defined as indigenous or informal. Yet African philanthropy is in fact the foundation on which an African's life and his or her development revolve. It is the foundation upon which modern institutions are built or from which they get their inspiration and identity. The bifurcation between informal and formal misses the central point about African societies; that one is an extension of the other." (Moyo 2011, 2)

It follows therefore that as a practice, philanthropy cuts through the ontologies and epistemologies of Africans in their various contexts. Most, if not all, Africans define themselves in relation to others, as opposed to the individualism that characterises other, particularly Western, societies. Although not the same as what we are describing as solidarity here, there is a strong resemblance between African solidarity and *minben*[2] in Confucianism. In Africa however, the most commonly-cited way to describe solidarity is one drawn from the Bantu languages, which in Zulu for example says *umuntu ngumuntu ngabantu*, literally meaning "*a person is a person because of people or through other people*". This is profound as it goes deep into the humanist elements of African philosophy. It is Afro-centric and pan-Africanist in thought and practice. The spirit of *ubuntu* engenders reciprocity and envelopes a communalism of interdependency, sharing, oneness, loving, giving, and a sense of a continuum of relationships. In other words, Africans see personhood as a process where one's humanity is affirmed by acknowledging others' humanity. Personhood is in essence about communities of interconnectedness; this sense has for the most part remained intact in most cultures, and where there have been ruptures as a consequence of the colonial encounter, a resurrection is underway simply because there is an understanding that community is a key building block of any society. Archbishop Desmond Tutu of South Africa captured this eloquently when he wrote about *ubuntu*:

> "It is the essence of being human. It speaks of the fact that my humanity is caught up and is inextricably bound up in yours. I am human because I belong. It speaks about wholeness, it speaks about compassion. A person with ubuntu is welcoming, hospitable, warm and generous, willing to share. Such people are open and available to others, willing to be vulnerable, affirming of others, do not feel threatened that others are able and good, for they have a proper self-assurance that comes from knowing that they belong in a greater whole. They know that they are diminished when others are humiliated, diminished when others are oppressed, diminished when others are treated as if they were less than who they are. The quality of ubuntu gives people resilience, enabling them to survive and emerge still human despite all efforts to dehumanize them." (Tutu 2005, 25–26)

The interdependence and linkages between development and the concept and practice of community cannot be better explained than this. Perhaps even more acutely, this interlinking and interdependence among various elements of an African – for example, between one's cultural systems and development, or one's giving and the common good – is best illustrated by Nelson Mandela, whose death in December of 2013 provided a platform for more giving and reflection on how the spirit of *ubuntu* can be used for development. Mandela embodied this philosophy and lived his entire life illustrating this concept, emerging as a legendary philanthropist. Not only did he give of his time, spending 27 years in prison for the liberation of his communities and society, he also went on to give of his money and kindness, and even established many institutions that are geared towards the realisation of the "spirit of *ubuntu*". Even at death Mandela still remained one of the greatest philanthropists using his will to distribute his estate to many from different spheres: schools, organisations, individuals, and his political party, the African National Congress (ANC). At his memorial service in Johannesburg, it was very clear why Mandela was and still is the greatest philanthropist, humanist, and pan-Africanist. There has never been a funeral as big as that of Mandela. One of his cousins captured Mandela's philanthropic character very well when he said:

> "Madiba shared his life with South Africa, Africa and the whole world. His life was about service to others. He mingled with kings, queens, prime ministers and ordinary people. He was the man of the people. A universal show of unity is a true reflection of what Madiba stood for." (General Thandoxolo Mandela speaking at Mandela's memorial service, 10 December 2013)

This article is about how this African philanthropy, with its similarities to pan-Africanism, can be utilised as the foundation and paradigm for transformational development in Africa. Mandela's life serves to illustrate the possibilities of utilising such a paradigm. For this reason, the article draws much from Mandela's life examples. There exist other examples as well; however, given the circumstances and context around which this article was written, there is no better example than that of Mandela. What he achieved through his selfless life is an indication of what is possible if Africa uses the values and principles enshrined in philanthropy and pan-Africanism to address developmental and governance questions. These are such features as risk-taking, community-mindedness, service, passion, humility, and solidarity among others. Africa cannot go wrong if it builds its developmental strategies and frameworks derived from these key foundational elements that are also central to the identity of an African. There is no nation or continent that has developed without recourse to its own cultures, so why would Africa be different? Nkosazana Zuma, the current Chair of the African Union, has prioritised African philanthropy in her strategy at the AU Commission by leading efforts to establish the AU foundation stated; at Mandela's funeral, she reminded us: *"Mandela represented solidarity – he lived these values and was always willing to serve. He was surrendered to the service of humanity"* (10 December, 2013).

It is clear that if African leaders surrendered themselves to the service of humanity, Africa would be well developed, well governed, and a number of conflicts would be eliminated. It is for this reason that we argue for philanthropy to be the paradigm for thought-leadership, as it requires development actors and processes to be at the service of humanity but at the same time to be resilient and absorb the challenges associated with development and transformation. This is what Mandela managed to do: he took risks and was imprisoned for his beliefs; he remained resilient even under immense suffering and spent his life serving humanity. Mandela taught and lived the single best lesson in development. President Obama summed this up very well when he said that *ubuntu* describes Mandela's life (in his statement at Mandela's memorial service, 10 December, 2013). Mandela achieved so much because of his embodiment of *ubuntu* and his teaching of others to find *ubuntu* in themselves. Mandela demonstrated that challenges to this vision can be overcome.

Writing this article, at this particular juncture in Africa's position in global relations and its own development, is significant in the sense that Africa has just recently celebrated 50 years of what many consider democratic governance since the demise of formal colonialism. It is also just over 50 years since the establishment of the Organisation of African Unity (OAU), now the AU, which is the foundation for pan-Africanist thinking – a worldview that emphasises solidarity, unity, and self-reliance among African member states. In these concepts, philanthropy and pan-Africanism are twins in developmental governance processes. Strikingly, the concepts of solidarity, unity, and self-reliance speak directly to the resolution of the challenges facing the continent. Imbedded in each of these are notions of accountability, empowerment, institution-building, and transparency – all key features of governance, which is an area that has vexed Africa for a long time but is crucial if the continent is to progress. This is also a time when Africa is poised to be the next growth centre, globally. Over the last few years, Africa's economic growth rates have averaged five per cent, and seven of the fastest developing economies are in Africa (Ethiopia, Mozambique, Tanzania, Congo, Ghana, Zambia, and Nigeria). This growth is not a result of the extractive sector alone (natural resources) but also of traditional sectors such as retail commerce, transportation, telecommunications, and manufacturing, among others. This has resulted in a growing middle class that has also begun giving back to communities through various philanthropic initiatives.

From a philanthropic point of view, this is the time to consolidate the momentum that has been generated over the last decades specifically for African philanthropy and its role in Africa's development. A number of factors therefore coalesce to make this period opportune for a discussion on African philanthropy and transformative development. Below, we show that African philanthropy is at the heart of Africa's development and as such ought to be *the* paradigm for any developmental frameworks and interventions. This is indeed embedded also in pan-Africanism, an ideology that continues to guide African thinking in general and the AU in particular. On the day of Mandela's passing, the lead author found himself listening to a recording of Mrs Graca Machel's address to the African Grant Makers Network, delivered in October 2012. As if to foreshadow this discussion, Machel told a story of how in 1993 she was invited by the Council on Foundations in the USA to talk about philanthropy in Africa. At that time, there were very few African philanthropic institutions that were recognised as such. Her own foundation, the Foundation for Community Development (FDC) was in its foundational phase and would be launched in 1994. At the meeting, she wanted to portray the differences in approach between African philanthropy and philanthropy from elsewhere. She told the meeting that in Africa, the starting point is a definition of a cause to be embraced, a cause that Africans feel passionate about; the second point is the generation of ideas to address the identified cause and definition of the scope of what is to be done. The final stage is the question of resources: where the resources will come from to address the cause. In other words, in African philanthropy, Mrs Machel argued:

> "money is not the first consideration but rather it is the cause that matters; yet elsewhere particularly in developed societies, in general money is first generated and in addition, due to legislation on how to spend money, the owners are somewhat obliged to then choose a cause to which they can give their money." (Machel 2012)

This story speaks directly to how as Africans, we first establish ourselves and our movements, and then go on to worry about financial resources. It does not mean that we do not give money in Africa – we do, but the motivation is first and foremost the cause, on which a chain of solidarity is built. This is how the pan-African movement also begun. We are concerned with how we engineer the ways of mobilising resources, time, knowledge, synergies, and of course money, for a cause. Machel's powerful story underlies the main point that the role of African philanthropy should be one of humility, synergies, and solidarity.

African philanthropy, pan-Africanism, and the quest for self-reliance

"Your neighbour's granary will never fill up yours." (Machel 2012)

One can infer these words from Graca Machel to mean that Africa cannot develop fully from foreign support, but rather that Africa needs to be self-sufficient and self-reliant in order to develop in transformative ways. This section traces the relationship between African philanthropy and pan-Africanism in their shared quest to make Africa self-sufficient and self-reliant. Although this article is not a treatise of pan-Africanism in detail, we want to draw the attention of the reader to the fact that both African philanthropy and pan-Africanism as approaches or ideologies have similar foundations and objectives. Both seek the self-reliance of humanity and socially just developmental outcomes. This is not the space to discuss the history of pan-Africanism; save to point out that it was in 1900 that Henry Sylvester-Williams, a Trinidad barrister, organised the first meeting of Africans and Africans of the Diaspora at the London Conference. Sylvester-Williams coined the term "pan-Africanism" for what had previously been called "the African movement". W.E.B. Du Bois, a promoter of the London Conference, applied the term "pan-African" to a series of six conferences that he convened in the capitals of European colonial empires from 1919 to 1945. As an intergovernmental movement, pan-Africanism was launched in 1958 with the first Conference of Independent African States in Accra, Ghana. Ghana and Liberia were the only sub-Saharan countries that were represented. Thereafter, as independence was achieved by more African states, other interpretations of pan-Africanism emerged, including: the Union of African States in 1960, the African States of the Casablanca Charter and the African and Malagasy Union in 1961, the Organisation of Inter-African and Malagasy States in 1962, and the African-Malagasy-Mauritius Common Organisation in 1964 (Nzewi 2008, 112).

For the purposes of discussing the functional dimensions of both African philanthropy and pan-Africanism, our attention here is on the formation of the Organisation of African Unity in 1963. The OAU was founded as a unifying pan-African institution, to continue the struggle for African liberation, integration, and socio-economic development. At the founding of the OAU in Addis Ababa, many speakers underlined what they saw as fundamental principles around Africa's development. The majority of these principles are values underpinning African philanthropy; in the main, these are unity, solidarity, and common purpose for the love of mankind. Speaking at the meeting, His Imperial Majesty Haile Selassie I, said:

> "Today we look to the future calmly, confidently and courageously. We look to the vision of an Africa not merely free but united. In facing this new challenge, we can take comfort and encouragement from the lessons of the past. We know that there are differences among us. Africans enjoy different cultures, distinctive values, and special attributes. But we also know that unity can be and has been attained among men [and women] of the most disparate origins, that difference of race, of religion, of culture, of tradition, are no insuperable obstacle to the coming together of peoples. History teaches us that unity is strength and cautions us to submerge and overcome our differences in the quest for common goals, to strive, with all our combined strength, for path to true African brotherhood [and sisterhood] and unity." (Selaisse 1963, 2)

He went on to emphasise the importance of unity which in philanthropic terms would be solidarity, and the other values enshrined in the spirit of *ubuntu*. He said:

> "Throughout all that has been said and written and done in these years, there runs a common theme. Unity is the accepted goal. We argue about techniques and tactics. But when semantics are stripped away, there is little argument among us. We are determined to create a union of Africans. In a very real sense, our continent is unmade, it still waits its creation and its creators." (Selassie 1963, 3)

The question is whether African philanthropy can be that creator of the next face of Africa. Selassie provided the framework and went on to give an example of how solidarity could be a tool for addressing Africa's challenges at the time. He identified the apartheid regime in South Africa, saying

> "We must redouble our efforts to banish this evil from our land. If we use the means available to us, South Africa's apartheid, just as colonialism, will shortly remain only as memory. If we pool our resources and use them well, this spectre will be banished forever." (Selassie 1963, 5)

This was echoed by various leaders as they went on to identify crisis points in Africa and argued for solidarity as a weapon to overcome them. It can be argued that during this time, these leaders were in solidarity against oppression and colonial forms.

The challenge today for philanthropy is to identify what solidarity is *for,* as opposed to what is *against*. At the time there was a clear relationship between African philanthropy and what the leaders wanted to achieve. His Majesty Mwami Mwambutsa IV, the King of Burundi, might as well have been describing the history of African philanthropy when he gave his speech:

> "The various African civilisations which preceded the colonial era resembled one another from various points of view. For example, the spirit of family solidarity was found everywhere, and the idea of hospitality was similarly general. Indeed, a careful study of the various African civilisations shows surprising similarities which make it clear that African unity is not a chimera-like and superficial construction, but a living entity which requires only to be translated on to the institutional plane." (Mwambutsa 1963)

The other reason that African philanthropy and pan-Africanism dovetail and speak so deeply to the question of solidarity and unity is that both are geared towards "love for humanity". Ahmadou Ahidjo, then President of Cameroon, captured this succinctly:

> "The voice of Africa has got to be heard, the voice which proclaims in appealing tones its love for mankind, which reminds us that the finest emotion on earth is not simply that aroused by the clash of arms." (Ahidjo 1963, 15)

In arguing for solidarity, unity, and relational existence, one could not have outlined the reasons more forcefully than Kwame N'krumah (1963, 34) of Ghana:

"No sporadic act or pious resolution can resolve our present problems. Nothing will be of avail, except the act of a united Africa."

There is no doubt therefore that pan-Africanism was seen by these leaders as a philosophy and an ethical system. As a philosophy, pan-Africanism represented the aggregation of the historical, cultural, spiritual, artistic, scientific, and philosophical legacies of Africans from past times to the present. As an ethical system, pan-Africanism promoted values that were the product of African civilisation and the struggles against slavery, racism, and colonialism (Hakim and Sherwood 2003). The pan-African movement evolved into a political entity with a clear agenda of eradicating all forms of oppression, slavery, and colonialism. It also sought to end racism, the dehumanising treatment of Africans, and aimed at political and socio-economic emancipation of Africa. Those who were involved in the movement included the likes of Marcus Garvey, W.E.B. Du Bois, Kwame Nkrumah, Henry Sylvester-Williams, Julius Nyerere, Jomo Kenyatta, and Nelson Mandela, to name just a few. One common thread among these individuals is their strong belief in solidarity, unity, and self-reliance. The story of Mandela, as told in his autobiography *The Long Walk to Freedom*, is very illustrative of this point. Not only did Mandela urge his colleagues and various political formations to be united against apartheid, he also sought assistance from other African leaders. This is why one of the victories the OAU can claim is the end of apartheid and the establishment of majority rule in South Africa. Today of course, pan-Africanism has evolved and can be seen in the efforts to promote even greater African economic, social, and political integration as spearheaded by the AU (Nzewi 2008, 112).

That Africans' humanity is bound to each other is a theme that dominates the spirit of *ubuntu* and pan-African thinking. This argument would not be complete without quoting extensively from one of the notable anthropologists of all times, Leopold Sedar Senghor of Senegal. He summed up this point aptly:

"Most of us feel that what brings us close to one another and must unite us is our position as underdeveloped countries, formerly colonised. Nor is that wrong. But we are not the only countries in that position. If fact could be said objectively to be whole truth, then African unity ought one day to dissolve with the disappearance of underdevelopment. I am convinced that what binds us lies deeper ... what binds us is beyond history. It arises from geography, ethnology, and hence from culture. It existed before Christianity and Islam; it is older than all colonization. It is that community of culture which I call African-ness. I would define it as 'the sum total of African civilised values'; African-ness always shows the same characteristics of passion in feelings, and vigour in expression. The consciousness of our community of culture, our African-ness, is a necessary preliminary to any progress along the road to unity." (Senghor 1963, 85)

An intended objective of pan-Africanist thinking was also that countries would be self-reliant and shake off dependence on the West. Hence the OAU Charter makes reference to the fact that all people have a right to control their own destiny. The preamble of the Charter states that the leaders are:

"... inspired by a common determination to promote understanding among our peoples and cooperation among our states in response to the aspirations of our peoples for brother-hood and solidarity, in a larger unity transcending ethnic and national differences."

This became one of the main objectives of the OAU, "*to promote the unity and solidarity of the African states as well as coordinate and intensify their cooperation and efforts to achieve a better life for the peoples of Africa*". Such better life could only be realised if Africa became self-reliant. For this reason, the *Monrovia Declaration of Commitment of Heads of State and Government, of the OAU on guidelines and measures for national and collective self-reliance in social and economic development* (1976) stressed the need:

> "[t]o ensure that member states individually and collectively restructure their economic and social strategies and programmes so as to achieve rapid socio-economic change and establish a solid domestic and intra-African base for a self-sustaining, self-reliant development and economic growth."

The Declaration went further to commit member states and their leaders to "*individually and collectively establish national, sub-regional and regional institutions that would facilitate the attainment of the objectives of self-reliance and self-sustainment*". It was very clear from this declaration that "*development policies ought to reflect adequately Africa's socio-cultural values in order to reinforce Africa's cultural identity*". The result was a series of declarations and frameworks that finally led to the Treaty establishing the African Economic Community which, among other issues, also stressed the need for solidarity and collective self-reliance. What runs through the entire treaty are the self-reliance and cooperation themes, captured very well by one of the primary objectives of the Treaty, which is:

> "[t]o promote economic, social and cultural development and the integration of African economies in order to increase economic self-reliance and promote endogenous and self-sustained development."
> (OAU 1991)

The concept of self-reliance is at the core of philanthropy. In Kenya and India, for example, self-reliance was utilised for development purposes, albeit with different objectives. In post-independence Kenya, the self-reliance initiative of *harambee* (meaning "let us all pull together") was used for rural development through the voluntary contribution of resources such as labour and cash (Waithima 2012, 5–6; Ngau 1987). *Harambee* was successful in the sense that over the period 1980–84, 12% of all national capital was generated through *harambee* while, by the end of 1980s, about 50% of all secondary schools were built through the initiative (Waithima 2012, 5–6). The strategy of *harambee* is still apparent in present-day Kenya, even though at some point it was politicised. We continue to see communities pulling together in Kenya in particular when there are disasters or humanitarian crises. The recent attack on the Westgate mall brought a number of Kenyans together and collectively raised many resources to respond to the catastrophe. The same was true after the 2007 election violence and the famine that faced some parts of Kenya. Similarly, in India in the 1900s, the *swadeshi* (meaning self-sufficient) movement was an economic strategy of the Indian citizens aimed at removing the British Empire from power, improving the economic conditions in the country, and promoting the notion of self-help, fellow-feeling, solidarity, and a sense of national identity (Radhakrishnan and Rao 2013, 1–2; Flanya 2013). The Négritude cultural movement also emphasised self-reliance, African self-determination, solidarity, and self–respect (Banoum 2011).

It can thus be concluded that the theme of self-reliance has been used over the years by individuals, movements, communities, and organisations for economic and political independence as well as self-independence. From being the overarching goal for individuals and movements to being identified as the blueprint for attaining independence and socio-economic development by communities and organisations, self-reliance is both a strategy and an ideology (Kim and Isma'il 2013, 586). It is also a key feature of philanthropy and pan-Africanism. The main point here is that different philanthropic practices are underpinned by the desire for self-reliance. This is the issue today that the AU and other pan-African institutions are trying to settle in the various initiatives that are aimed at developing Africa. Below is a short discussion of some of those processes.

Self-reliance and self-sufficiency today

Today a number of frameworks and initiatives are in place precisely to address the twin questions of self-reliance and self-sufficiency. Most of these are at the pan-African level while others are at the national and regional levels. To put things into perspective, and summarise where we are today: our reference is the year 2013 and the activities that caught the attention of many. The year 2013 marked the 50th anniversary of the founding of the OAU, and was greeted by many celebrations and reflections. The 21st AU Summit was held in Addis Ababa in May 2013, and the theme centred on Pan-Africanism and the African Renaissance. Key decisions from this Summit that are relevant to this article include:

- the creation of an AU Foundation for voluntary contributions towards financing the AU;
- further commitment to gender mainstreaming and youth empowerment;
- an African common position on the Post-2015 Development Agenda;
- an effective implementation of the Roadmap on Shared Responsibility and Global Solidarity for AIDS, Tuberculosis, and Malaria;
- the establishment of an African Capacity for Immediate Response to Crises and the substantial increase of contributions to the Peace Fund; and
- the need to build innovative, flexible, action-oriented, and balanced international partnerships.

This was at the backdrop of a continental vision that the AU has developed – Agenda 2063 – that seeks to envision what Africa would be in 50 years: a continent free from poverty and conflict, where socio-economic development is driven by its own citizens. This of course cannot happen unless the twin concepts of self-reliance and self-sufficiency are central in this kind of visioning. The Draft Framework Document outlining this agenda states that:

> "Agenda 2063 should be seen as an opportunity to recreate the African narrative by putting to perspective to enthuse and energise the African population and use their constructive energy to set and implement an achievable agenda for unity, peace and development in the 21st century. The thrust of the Agenda 2063 is a programme of social, economic and political rejuvenation that links the past, present and the future in order to create a new generation of Pan-Africanists that will harness the lessons learnt and use them as building blocks to consolidate the hope and promises of the founding parents for a true renaissance of Africa." (AU 2013a, 15)

Just like the project of pan-Africanism, Agenda 2063 might fail without political leadership. There is also a danger that this too might be shelved if it is not grounded on Africa's peoples. It might suffer the fate of such great initiatives as the Abuja Treaty, the Lagos Plan of Action, and the Final Act of Lagos. There is also a danger of a lack of resources to finance the implementation of this agenda. For this reason, it is important that the Agenda also includes local mobilisation of resources. There is no doubt that Agenda 2063 is an aspiration that is also captured in the consultations around the post-2015 MDGs framework. What has emerged particularly from the African continent is a realisation that for Africa to progress, its economies must be transformed structurally. However, there is recognition too that economic growth alone will not lead to inclusive development.

For this reason, the agenda of the continent also includes addressing conflicts and wars. To do so, once again the AU has drawn lessons from African philanthropy and established an initiative that builds on solidarity. The African Solidarity Initiative (ASI) is a programme of the Peace and Security department of the AU dedicated to resolving conflicts on the continent by utilising social capital and other solidarity principles. With its core message as "Africa helping Africa", this initiative was launched at the 19th Ordinary Session of the Policy Organs of the Union in July 2012. Its main mandate is to mobilise support from within the continent for countries emerging from conflict in line with the AU policy on Post-Conflict Reconstruction and Development. This initiative has been described by the AU as:

> "a process intended to harness and expand the spirit of African solidarity and self-help to support on-going efforts on post conflict reconstruction and development in a number of African countries." (see pamphlet by the AU on ASI)

Many of the principles underpinning this initiative are similar to those of African philanthropy. For example, the main objectives of this initiative include:

- deepening the essence of African solidarity and promote a paradigm shift which centre-stages African mutual assistance as a key dimension for enhanced and effective development of the continent; and
- providing a unique opportunity for generating additional "outside the box" ideas for addressing PCRD challenges, by actively involving African countries, relevant organisations/institutions, parastatal, private sector, philanthropy organisations/foundations, academia, civil society, faith-based organisations, African experts, and the Diaspora.

When launching ASI, the ministers of foreign affairs underlined the importance of solidarity in the achievement of peace and development. They said:

> "Our objective is to promote African solidarity, mutual assistance and regional integration; and propel the continent to a higher level of development and self-confidence; driven by the motto: 'Africa helping Africa'." (Declaration on the Launch of the African Solidarity Initiative [ASI] for the Mobilisation of Support for Post Conflict Reconstruction and Development in Africa)

Speaking at the Africa Solidarity Initiative conference at the AU Commission in February 2014, President Jacob Zuma of South Africa made the point that this idea of African solidarity is not new but was utilised during the liberation struggle. As such it is important that African countries are in solidarity with those emerging from conflict. He further underscored the point that African solidarity is rooted in African culture and pan-Africanism. At this conference, a number of countries pledged their support for ASI in general and the African Union Mission in the Central African Republic in particular. For example, Ethiopia pledged US$500,000; South Africa pledged US$1 million and political support; Nigeria pledged US$3.5 million and technical support; Algeria pledged US$1 million and airlifting of troops; Cote d'Ivoire pledged US$500,000, and Gambia pledged US$50,000. There were pledges also from Europe, Asia, and the Americas. In many ways this underscored the closeness and interdependence between African philanthropy and development. In African cultures one's neighbour cannot go hungry when the other could give help; this is reflected in African philanthropy. Giving and sharing is at the heart of philanthropy. This is precisely what ASI is promoting. The challenge of course will be to make this home-grown and driven by Africans themselves.

The other continental initiative that is worth discussing in this context is the African Union Foundation, a new development in Africa. The AU Foundation was established through an Assembly Decision which created it as a vehicle for voluntary contributions towards financing the African Union (see Decision on the Establishment of an African Union Foundation for Voluntary Contributions towards Financing the African Union; Doc.Assembly/AU/6 (XX1)). This occurred against the backdrop of the High-Level Panel on Alternative Sources of Financing for the African Union that was headed by former President of Nigeria, Olusegun Obasanjo. In May 2013, Obasanjo submitted his report which recommended five options:

- Private sector funding
- Levy on insurance premiums (rate of 1%)
- Levy on international travel (US$2.5 for travel outside the continent and US$1 for travel within the continent)
- Tourism and hospitality (US$1 for each hotel stay)
- Import levy (0.2% on goods imported from outside the continent)

The recommendation to establish a foundation was made as part of the framework to hold donations from the private sector and other contributions. In briefing potential Foundation council members at the AU Commission on 1 February 2014, the AU Chairperson Nkosazana Dlamini-Zuma said:

"The foundation is one of the mechanisms to help Africa take charge of its own affairs. Africa is very rich but Africans are very poor, and this has to change, be turned around to eradicate poverty. Poverty is not an accident; it is man-made and can be removed by the actions of people." (Quoting Mandela; notes taken by the lead author, who attended the meeting in Addis Ababa)

She went on to state that the foundation will prioritise funding, especially for the Agenda 2063 pillars, such as human development, youth development and entrepreneurship, regional integration, women and gender equality, and management of diversity. In this regard she said the mission of the foundation is to:

"[m]obilize resources in support of the African Union's vision of an integrated, people-centered and prosperous Africa, at peace with itself and taking its rightful place in the world ... It is time for Africa to mobilize our own resources in support of our development and take charge of our own destiny." (See Press Release No.40/22nd AU Summit: Newly Established 'African Union Foundation' Holds Inaugural Promoters' Meeting in Addis Ababa, 1 February 2014)

In many ways, the foundation will be catalytic in mobilising Africans for an African agenda for development. As such it must not be seen only in financial terms but rather as a mobilisation tool for active participation in affairs affecting Africans. Dlamini-Zuma summarised this very well:

"The Foundation will strive to more deeply engage Africa's private sector, African individuals and communities, and leading African philanthropists to generate resources and provide valuable insight on ways in which their success can accelerate Africa's development. The issue of domestic and alternative sources of funding has been an intrinsic element of the continent's commitments of the Pan-African values of self-determination, solidarity and self-reliance." (Press Release No. 40/22nd AU Summit: Newly Established 'African Union Foundation' Holds Inaugural Promoters' Meeting in Addis Ababa, 1 February 2014)

These and other initiatives such as the African Grant Makers Network, the Africa 50 Fund, and other already-existing philanthropic initiatives are aimed at addressing the many challenges facing Africa such as poverty, a heavy disease burden, high rates of unemployment, rampant corruption and conflicts, weak intra-African trade, and slow integration. In the midst of these problems and the pessimism and doubt that exist amongst African leaders and their citizens, is it possible for the continent to reduce and ultimately end its excessive dependence on outside forces to provide aid for development? We maintain that it is possible, especially if the above initiatives and others are aggressively implemented. African countries ultimately need to bear the primary responsibility for the continent's challenges; Africa is the major driver of its own renaissance through self-determination and self-reliance. The opportunities for African partnership and strategies which focus on solutions to African problems exist and are becoming a must, as discussed above. In recent years, a number of developments in African countries have taken place (and still occur presently) which prove that African self-reliance and self-determination are possible. The continent was dubbed "hopeless" on the cover of the May 2000 issue of *The Economist*, but this is a stigma that Africa is slowly but surely shedding. Eleven years later, on the cover of the December 2011 issue, *The Economist* hailed Africa as "the hopeful continent".

Today, the African continent is one of the fastest-growing and emerging economic regions in the world. The continent has clearly realised that it has to take its destiny into its own hands and that it has to draw upon its own resources to achieve sustainable development. There has been an increasing awareness of the need to integrate economically, socially, and environmentally in order to achieve sustainable development. Moreover, there is a recognition that improved access to education (especially for women), market liberalisation and privatisation, democratisation, and increased regional and sub-regional economic integration and cooperation are important for progress and self-reliance. This is happening in Africa. According to the World Bank's "Africa's Pulse" report, most African economies are flourishing, and the GDP growth rate in Africa could reach 5.3% in 2014 and 5.5% in 2015 (World Bank 2013, 2). The report maintains that this is due to strong government interventions such as investments, and higher production in the mineral resources, agriculture, and service sectors.

Through initiatives such as the African Union Roadmap on Shared Responsibility and Global Solidarity (2012–15), the fight against AIDS, TB, and malaria in Africa has met with some prominent success. The initiative, which urges African countries to increase domestic funding for health and decrease their reliance on external donor funding, has resulted in at least seven million people receiving HIV treatment across Africa, with nearly one million added in the last year. Additionally, new HIV infections and deaths from AIDS continue to fall (UNAIDS Press Release 2013). This is due to strong leadership and shared responsibility in Africa working together and looking within for solutions to combat these diseases.

Other improvements that the continent has made are marked by the fact that in general a number of countries are at peace even though conflicts still persist in North Africa and some parts of sub-Saharan Africa (in Somalia, Central African Republic, and Sudan, among others); education is being prioritised and more children are attending school; and life expectancy has risen by a tenth in the last decade. Also, consumer spending will almost double in the next 10 years, modern technology is being embraced and used to increase productivity, and citizens of African countries are voting and trying to hold their governments accountable.

While many will argue that these improvements are to a large extent due to outside influences and foreign aid, the main saviours of the continent have been its own people and the active role that they are undertaking to make progress in their countries. The citizens of Africa are calling for better leadership and governance, they are embracing self-determination and innovation, and they are using their own resources to make progress. No doubt, much of this is due to the rise in commercial activities like remittances by Africa's Diaspora and African philanthropy-factors that are driving the process of reducing the continent's dependence on external aid and promoting self-reliance.

While the progress in independent Africa should be applauded, it is important to note that Africa's transformation is still incomplete. The continent has a long way to go to be free from poverty, disease, and conflict and to achieve full self-reliance. But the potential and opportunities exist if only we harness the power of African philanthropy and its underpinning values that are also key features of pan-Africanism. Equally crucial are strategies to address some of the challenges that might arise in the process such as lack of political will, inadequate funding, competing agendas, and corruption, as we have seen in the past with similar progressive initiatives. However, embedded within the African philanthropy framework are questions of governance, transparency, and accountability, for how can one be in solidarity with others unless they are accountable. This should be the foregrounding for any developmental initiative.

Conclusion

The central argument of this article is that African philanthropy, by its very definition and practice, is and ought to be the foundation upon which transformational development takes place in the continent. Sharing its values, foundations, and premises with pan-Africanism, African philanthropy as an achievable development paradigm was illustrated in our lifetime by no other than Nelson Mandela. Mandela's philanthropic work epitomised solidarity, interconnectedness, interdependencies, reciprocity, mutuality, loving, sharing, passion, humility, risk-taking, and a continuum of relationships. Embedded in each of these are notions of accountability, empowerment, institution-building, and transparency – all key features of governance and all key features for development. This is what African philanthropy is about: surrendering oneself to the service of humanity while being resilient and absorbing the challenges associated with development and transformation.

The article also illustrates that self-sufficiency and self-reliance are key features of philanthropy and pan-Africanism. Both are central to the envisioning of a number of frameworks and initiatives in Africa today such as Agenda 2063 and the African Solidarity Initiative. Also, as an ideology and a strategy, both are possible in Africa. African countries and the people of Africa, in the past and at present, have looked towards their own resources and capabilities for development. That said, if African philanthropy exudes values such as self-reliance, self-sufficiency, interconnectedness, interdependencies, reciprocity, accountability, and empowerment, African philanthropy then ought to be the foundation upon which transformational development takes place in the continent. In other words, it is possible for African philanthropy to be an anchor and foundation for development despite all the challenges thrown against it. For in the words of J. E. Casely Hayford (quoted in Owusu 1992, 379), *"our system being a communal one, it is a case of sink or swim with the family and the community… educated or uneducated, we sink or swim with our people"*.

Notes

1. In the previous articles, the lead writer has also focused specifically on the distinction between African philanthropy, philanthropy in Africa, and philanthropy with African features.
2. *Minben* is a doctrine that requires the government to treat the welfare of the common people as the foundation of its wealth and power; see the discussion by Shi and Lu 2010.

References

African Union. 2013a. *African Union Agenda 2063: The Future We Want for Africa* (Draft Framework Document, 6 December). Addis Ababa: African Union.

African Union. 2013b. *Celebrating Success: Africa's Voice over 50 years, 1963–2013*. Addis Ababa: African Union.

African Union. 2014. "Newly Established "African Union Foundation" Holds Inaugural Promoters' Meeting in Addis Ababa." Accessed July 31, 2014. http://summits.au.int/en/22ndsummit/events/newly-established-%E2%80%9Cafrican-union-foundation%E2%80%9D-holds-inaugural-promoters%E2%80%99-meeting-ad

Ahidjo, A. 1963. Speech to the OAU, in *Celebrating Success: Africa's Voice over 50 years, 1963–2013*, edited by the African Union. Addis Ababa: African Union.

Banoum, B. N. 2011. "Négritude." Accessed November 13, 2013. http://exhibitions.nypl.org/africanaage/essay-negritude.html.

Flanya, S. K. 2013. "Recourse to History- Reviving the Return-to –Africa Initiative." *AU Echo* 5: 1–11.

Kim, I., and M. Isma'il. 2013. "Self-Reliance: Key to Sustainable Rural Development in Nigeria." *ARPN Journal of Science and Technology* 3 (6): 585–591.

Hakim, A., and M. Sherwood. 2003. *Pan-African History: Political Figures from Africa and the Diaspora since 1787*. London: Routledge. 224.

Machel, G. 2012. "Address to the Africa Grantmakers Network Assembly," October 29, Johannesburg. Accessed July 31, 2014. http://www.youtube.com/watch?v=FS88Bn2Lx_k

Moyo, B. 2009a. "Philanthropy in Africa." In *International Encyclopedia of Civil Society*, edited by Helmut K. Anheier and Stefan Toepler, 1187–1192. New York, NY: Springer.

Moyo, B. 2009b. "Establishing the African Grant Makers Network." *Discussion Document Prepared for the Launch of AGN, Accra*.

Moyo, B. 2010. "Philanthropy in Africa: Functions, Status, Challenges and Opportunities." In *Global Philanthropy*, edited by N. MacDonald, and L. de Borms, 259–270. London: MF Publishing.

Moyo, B. 2011. *Transformative Innovations in African Philanthropy*. Brighton: IDS for The Bellagio Initiative. Accessed June 25, 2014. http://www.bellagioinitiative.org/wp-content/uploads/2011/10/Bellagio-Moyo.pdf

Moyo, B. 2013. "Trends, Innovations and Partnerships for Development in African Philanthropy." In *Giving to Help, Helping to Give: The Context and Politics of African Philanthropy*, 37–63. Dakar: Amalion Press.

Moyo, B., and T. Aina. 2013. *Giving to Help, Helping to Give: The Context and Politics of African Philanthropy*. Dakar: Amalion Press.

Mwambutsa, M. 1963. *Speech to the OAU, in Celebrating Success: Africa's Voice over 50 years, 1963–2013, edited by the African Union*. Addis Ababa: African Union.

Ngau, P. M. 1987. "Tensions in Empowerment: The Experience of Harambee (self-help) Movement in Kenya." *Economic Development and Cultural Change* 35 (3): 523–538.

N'krumah, K. 1963. Speech to OAU, in *Celebrating Success: Africa's Voice over 50 years, 1963–2013*, edited by the African Union. Addis Ababa: African Union.

Nzewi, O. I. 2008. "The Role of the Pan-African Parliament in African Regionalism (2004–2006): An Institutional Perspective." PhD diss., University of Pretoria.

Owusu, M. 1992. "Democracy and Africa-A View from the Village." *The Journal of Modern African Studies* 30 (3): 369–396.

OAU. 1976. "Monrovia Declaration of Heads of State and Government of the OAU on Guidelines and Measures for National and Collective Self-reliance in social and economic development." Accessed July 31, 2014. http://www.au.int/en/sites/default/files/OAU_Charter_1963_0.pdf

OAU. 1991. "Treaty establishing the African Economic Community." Accessed July 31, 2014. http://www.au.int/en/sites/default/files/TREATY_ESTABLISHING_THE_AFRICAN_ECONOMIC_COMMUNITY.pdf

Radhakrishnan, S., and V. K. R. V. Rao. "Swadeshi: Meaning and Contemporary Relevance." Accessed November 18, 2013. http://www.hss.iitb.ac.in/bhole/eog/ch14.pdf

Selaisse, H. 1963. "Speech to OAU." In *Celebrating Success: Africa's Voice over 50 years, 1963–2013*, edited by the African Union. Addis Ababa: African Union.

Senghor, L. 1963. "Speech to OAU." In *Celebrating Success: Africa's Voice over 50 years, 1963–2013*, edited by the African Union. Addis Ababa: African Union.

Shi, T., and J. Lu. 2010. "The Shadow of Confucianism." *Journal of Democracy* VL21 (4): 123–130.

Tutu, D. 2005. *God Has A Dream: A Vision of Hope for Our Times*. New York, NY: Image Books, Doubleday.

UNAIDS. 2013. "Press Release." Accessed November 27. http://www.unaids.org/en/resources/presscentre/pressreleaseandstatementarchive/2013/may/20130521prupdateafrica/

Waithima, A. K. 2012. "The Role of Harambee Contributions in Corruption: Experimental Evidence from Kenya." *Investment Climate and Business Environment Research Fund*, 5–39.

World Bank. 2013. "An Analysis of Issues Shaping Africa's Economic Future." Africa Pulse Report Accessed November 27, 2013. http://www.worldbank.org/content/dam/Worldbank/document/Africa/Report/Africas-Pulse-brochure_Vol7.pdf

Wiki approaches to wicked problems: considering African traditions in innovative collaborative approaches

Dawn S. Booker

Wicked problems are complex problems that are seemingly impossible to solve. However, an analysis of selected traditional African philosophies provides insight into how certain traditions may be applied in a practical sense to address social and environmental problems. Further, many newer collaborative and 'wiki'-based solutions provide a natural way for Africans and other global actors to participate in lessening the impact of global wicked problems. Ushahidi and the Geo-Wiki Project serve as examples of organisations that have provided a platform for this type of open development.

Les problèmes « épineux » sont des problèmes complexes qui semblent impossibles à résoudre. Cependant, une analyse de philosophies africaines traditionnelles sélectionnées donne une idée de la manière dont certaines traditions peuvent être appliquées de façon pratique pour aborder les problèmes sociaux et environnementaux. De plus, de nombreuses solutions collaboratives et basées sur des « wiki » plus nouvelles donnent aux Africains et à d'autres acteurs mondiaux un moyen naturel de participer à l'atténuation de l'impact des problèmes épineux mondiaux. Ushahidi et le Geo-Wiki Project constituent des exemples d'organisations qui ont fourni une plateforme pour ce type de développement ouvert.

La denominación de problemas malvados hace referencia a problemas complejos que parecen imposibles de solucionar. Sin embargo, un análisis de las filosofías africanas tradicionales ofrece elementos en torno a cómo ciertas tradiciones pueden ser aplicadas para abordar los problemas sociales y medioambientales desde un punto de vista práctico. Por otra parte, muchas de las nuevas soluciones colaborativas y de tipo «Wiki» brindan modos naturales a partir de los cuales los africanos y otros actores a nivel global pueden participar para mitigar el impacto de los problemas malvados a nivel mundial. Tanto Ushahidi como el Proyecto Geo-Wiki representan ejemplos de organizaciones que han creado plataformas para este tipo de desarrollo abierto.

Introduction

The term wiki seems to have many meanings and derivations. The word means "quick" or "fast" in Hawaiian. Wiki also appears as an acronym, meaning "what I know is". The word has come to denote collaborative efforts in which people come together to pool their knowledge and expertise with a minimum of fuss and formality. There is a large and growing literature on wiki principles with different, contending camps and positions. Wiki also refers the collaborative way that data is collected and referenced online. In essence, a wiki is a content-management system without defined owners or leaders and that has little implicit structure, which emerges according to the needs of the users (Mitchell 2008).

A closer examination of practical applications of the term wiki provides insight into how this collaborative approach may be used to solve or lessen the impact of complex global issues. These social and environmental issues are often referred to as "wicked problems". Wicked problems are considered difficult or impossible to solve for several reasons, including incomplete or contradictory knowledge, the number of people and opinions involved, the large economic burden, and the interconnected nature of these problems with other problems. This paper explores the conditions of wicked problems and the ways in which a wiki approach can be practically used to foster co-innovation and collaboration to address global issues.

Specifically, based on a broader definition of the term wiki, it seems that the collaborative and interconnected nature of a wiki approach aligns with particular African philosophies including the concept of *ubuntu*, or a shared humanity. It is important to note that there is not an implication of or reference to a single philosophy or set of philosophies that applies to the entire continent. However, my research suggests that there are threads that run through very many African countries and regions. Further, an analysis of the conditions of social and environmental wicked problems, while at the same time considering the traditions of community and collective approaches often present in African communities, leads to a model of a wiki approach to wicked problems. Considering wicked problems from a transdisciplinary research perspective provides the foundation for seeking innovative ways to approach these complex and often seemingly unsolvable problems.[1]

These considerations lead to the following research question: *"Does the overlap in discourse within wiki approaches, wicked problems, and some traditional African philosophies provide a practical and natural way for Africans and other global actors to co-innovate to address wicked social problems?"*

This article reviews existing literature to define what is meant by wiki and to explain a "wiki approach". The concept of wicked problems is explored and defined in social and environmental contexts. An analysis of what is meant by traditional African philosophies and which traditional concepts are representative of this philosophy provides insight into how certain traditions may be applied in a practical sense. After examining the existing literature on these three principles, the balance of the paper focuses on the details and implications of the overlap in discourse. The open-source technology company Ushahidi and the Geo-Wiki Project serve as examples of organisations that successfully apply this intersection to address wicked problems in Africa. Finally, the implications of this research are applied to suggestions and recommendations for practitioners and global actors tasked with addressing the world's wicked problems.

Wiki/wiki approach

The idea that local culture and traditions should be considered in approaches to development is not new; however, applying similar discourse markers that are present in both the problem and in the affected culture is conceptually unique. A discourse analysis of the existing literature on wicked problems, collaborative approaches, and different traditional African philosophies was conducted to find similar discourse markers. Academic and contemporary business literature are identified and analysed to arrive at a model for a wiki approach.

William Hitt, in his book titled *The Global Citizen*, says addressing global wicked problems requires a new way of thinking. Hitt (1998) argues that most life-and-death problems facing humanity are global and will not be resolved by individual nation-states working independently. The only way that humanity can cope is through building a global community. Thus wiki approaches to wicked global problems may provide a platform for the global cooperation necessary to address these problems.

Hitt (1998) goes on to say that identities shape interests, which can be interpreted to mean that for any approach to wicked problems to be successful, identity must be considered. Many sub-Saharan Africans consider their shared traditions and philosophies to be an integral component of their collective identity. The interconnected nature of the world's wicked problems creates ideal conditions for the application of aspects of local culture and traditions to address local and regional problems. Examining the intersecting discourse on wiki approaches, wicked problems, and traditional African philosophies may provide a basis to lessen the global impact of these issues. This discourse overlaps around the ideas of community and collective approaches in that very often traditional African philosophies and wiki approaches include ideas of community and collaboration. These ideas are often present in the discourse that addresses the best practices in addressing wicked problems.

The word *wiki* has also been used to represent collaborative online efforts to share collective intelligence. A wiki is a website that allows users to add, remove, or edit content quickly and easily. This ease of interaction and operation makes a wiki an effective tool for collaborative production. The term wiki can also refer to the collaborative software itself (wiki engine) that facilitates the operation of such a website. A wiki is a content-management system that differs from blogs or most other such systems in that the content does not have a defined owner or leader. Wikis tend to have limited implicit structure, allowing the structure to emerge according to the needs of the users (Bryant 2006). An obvious example of a wiki would be the website *Wikipedia*.

However, for the purpose of this paper, a slightly broader definition of the term, based on contemporary business literature, is used. Tapscott and Williams coined the term "wikinomics" in 2006 in a popular business book titled *Wikinomics: How Mass Collaboration Changes Everything*. According to Tapscott and Williams (2006, 36), "*as the new Web and the Net Gen [eration] collide with the forces of globalization, we enter what might be considered a perfect storm, or an instance in which converging waves of change and innovation are toppling conventional economic wisdom*". They argue that the new promise of collaboration is that we will have the ability to efficiently harness human skill, ingenuity, and intelligence through peer production. Collaborating in this way allows for the integration of the talents of dispersed individuals and organisations to become the defining competency for managers and firms. Existing business literature suggests that exploring the practical applications for the term "wiki" allows for a broader definition that could be considered a wiki approach (Shirky 2008). The following key terms or markers are prevalent in the literature that explores collaborative approaches to solving problems.

Collective knowledge

In his book *Leadership in a Wiki World: Leveraging Collective Knowledge to Make the Leap to Extraordinary Performance*, Collins (2011) states that the smartest organisations are those with the capacity to access the wisdom of the crowd. Further, in instances in which organisations have the processes to leverage collective knowledge, nobody is smarter or faster than everybody. Therefore, accessing the collective knowledge of the entire organisation challenges the idea that excellence comes from individual knowledge of heroic leaders and star performers.

Collective action

Meinzen-Dick and DiGregorio (2004) define collective action as a voluntary action of a group to achieve common interests. Members can act directly on their own or through an organisation.

Collective action is increasingly recognised as a positive force for improving lives in the developing world, especially among the poor (Grootaert 2001). Collective action occurs when people join to tackle problems of common interest and can be an effective means of group problem solving, especially among poorer residents of densely populated urban or farming areas (Coppock and Desta 2013).

Sharing

In the book *Share or Die*, Harris (2012) indicates that the ways to share in everyday life seem to be multiplying like rabbits. He observes that the Great Recession may be responsible for forcing all of us to pay more attention to opportunities and avenues for sharing these days. The so-called "sharing economy" includes car sharing, ride sharing, bike sharing, yard sharing, Massive Open Online Courses (MOOCs), co-working, co-housing, tool libraries, and all kinds of cooperatives that encourage sharing for problem solving and sustainability. There are also ways to share power, dialogue, and knowledge, such as workplace democracy; citizens' deliberative councils; "un-conferences" of open, self-organised gatherings; and "world cafés" for focused deliberations. According to Harris, we are using 50% more natural resources per year than the earth can replace, and the global population and per capita consumption are increasing. It is now glaringly obvious that we need to learn to share on a global scale – fast – or die. There are also several start-ups that are helping people meet real needs. These include Airbnb (peer accommodations), Thredup (kids' clothes), Velib (bicycle sharing), Chegg (textbook rentals), Neighborgoods (general sharing), RelayRides (peer-to-peer car sharing), Hyperlocavore (garden dating), Zimride (ride sharing), and many others (Harris and Gorenflo 2012).

Peer production

Benkler (2003) describes peer production as a process by which many individuals contribute to a joint effort that efficiently produces a unit of information or culture. Neither managers nor price signals in the market coordinate the actions of these individuals. Benkler shares that in peer production, communities and individuals band together to contribute to things they care about by volunteering their time, talent, and effort in small or large ways. He stresses that these efforts mark the appearance of a new mode of production, one that was mostly unavailable to people in the physical economy, with the exception of traditional collective community activities (Benkler and Nissenbaum 2006). An early examination of peer production appears in the writing of Raymond (1999), who contrasts the openness of Linux's code to the hierarchical and closed way of most software released prior to Linux. The bazaar, he says, is open and available to all, in contrast with the cathedral, which is more closed and restricts access to the code to a privileged few. Open and available access to tools and modes of production can fundamentally alter the producer/consumer relationship with regard to culture, entertainment, and information. In a sense, an open environment levels the playing field as a consumer becomes a producer by gaining access to the tools of production (Benkler 2003).

Collaboration

Collaboration is about deliberately creating interactions out of which, we hope, something positive will emerge. The notion of collaboration is about establishing a place to find common ground out of which we can create new opportunities (Schultz 2013). Additionally, the "culture of contest" that results from competition is becoming increasingly maladaptive in an age of ever-increasing social and ecological interdependence (Karlberg 2004). Furthermore, the most valuable features of competition – the pursuit of excellence, innovation, and productivity – are not merely contingent on self-interested behaviours, which result in winners or losers. On the contrary, these desired outcomes of competition assume their most mature form within the framework of cooperation and mutual gains – a framework of collaboration. Often collaboration can still be executed with a culture of contest; however, collaborative efforts that are effective in addressing global issues must be more organic in nature, allowing collaboration to be motivated by concern for the entire social body (Khozein, Karlberg, and Freeman 2013).

The aforementioned principals represent components of a wiki approach in that they incorporate the ideas of informal and collective production. A wiki approach is less about top down management and hierarchy and more about solutions that arise from within a community or through the input and interaction of many. There is an emphasis on access and openness in environment where power and knowledge are readily shared.

Wicked social and environmental problems

Whenever interest groups with strongly divergent values are well organised and highly motivated, and when uncertainties in the science surrounding an issue may be exploited, an issue can move into the realm of a wicked problem (Balint 2011, 1–5). Climate change, the HIV/AIDS epidemic, world hunger, and determining sustainable uses of natural resources are all examples of wicked problems. Wicked problems are complex issues that resist conventional problem-solving approaches and for which existing solutions often create unintended consequences that only make the original problem worse (Rittel and Webber 1973). The concept of wicked problems originates in social planning theory and describes complex, interconnected issues. The term wicked does not denote "evil", but serves as an indication of the complexity of these problems. Additionally, the unstructured nature of wicked problems makes them extremely difficult to identify, which further adds complexity and uncertainty (Weber and Khademian 2008). Wicked problems are differentiated from tame problems according to the following conditions of global wicked problems:

- There is incomplete or contradictory knowledge about the problem.
- Addressing wicked problems involves the opinions of many stakeholders.
- Often, with a wicked social-environmental problem, the problem itself is in the eye of the beholder, or the stakeholder.
- There is no single correct formulation of the problem.
- Wicked problems typically involve a large economic burden.
- The interconnected nature of these problems with other problems means that changes to one wicked problem can affect another (Rittel and Webber 1973).

Consequently, there is no single, correct, optimal solution. The decision maker must come to a conclusion without knowing if all feasible and desirable options have been explored, and any management choice will ultimately be better or worse rather than true or false.

Economist Vijay Vaitheeswaran (2012) addresses new and innovative approaches to wicked problems in the book *Need, Speed, and Greed: How the New Rules of Innovation Can Transform Businesses, Propel Nations to Greatness, and Tame the World's Most Wicked Problems*. He notes that despite the best intentions of central planners, top down innovation does not work as well as the bottom-up variety. He argues that a collaborative approach to tackling global issues is the best way to address global wicked problems. He says that dealing with global wicked problems will be difficult, but if governments partner with the private sector, the collaboration can help solve issues of governance and market failures. These problems can be addressed if we harness them and redirect the creative energies of entrepreneurs and corporations to take on – and profit from – solving the world's wicked problems rather than wishing them away or demonising them (Vaitheeswaran 2012).

Traditional African philosophy

What is meant by traditional African philosophy? In attempting to offer a definition, I would like to acknowledge that the use of the word "African" in no way insinuates that there is a monolithic African philosophy or one single set of traditions that apply to the entire continent. It is understood that there are 54 African countries with different and connected traditions, heritages, and cultures. However, there is evidence that there are threads that run through very many African countries and regions. For decades, African scholars outside the social sciences have consistently claimed that there have been, are, and will continue to be widespread psychological and cultural themes and patterns that are unique to sub-Saharan Africa (Lassiter 2000).

The idea of interdependence and interconnection is often cited as a cultural theme expressed in many distinct terms and phrases throughout the region. South African philosophy professor Augustine Shutte (1993), citing the Xhosa proverb *umuntu ngumuntu ngabantu* (a person is a person through persons), writes:

"This [proverb] is the Xhosa expression of a notion that is common to all African languages and traditional cultures ... [It] is concerned both with the peculiar interdependence of persons on others for the exercise, development and fulfilment of their powers that are recognized in African traditional thought, and also with the understanding of what it is to be a person that underlies this ... In European philosophy of whatever kind, the self is always envisaged as something 'inside' a person, or at least as a kind of container of mental properties and powers. In African thought, it is seen as 'outside,' subsisting in relationship to what is other, the natural and social environment. In fact the sharp distinction between self and world, a self that controls and changes the world and is in some sense 'above' it, this distinction so characteristic of European philosophy, disappears. Self and world are united and intermingle in a web of reciprocal relations." (1998, 46–47)

The concept of community or communalism is an important theme often found in sub-Saharan Africa. However, it is important in a cross-cultural as well as within a multicultural context to distinguish what is meant by communal and community. Although communal values seem to be a polar opposite to what is often described as a Western value of individualism, it is possible for an individual to maintain his/her own identity along with establishing a potential creative role in a community (Bell 2002).

Although there is diversity and a vast number of subcultures in Africa, there is a foundation of shared values, attitudes, and institutions that bind the nations of sub-Saharan Africa. Additionally, throughout the continent, the bond between religion, tradition, and society remains strong (Manguelle 2000). Some additional threads of tradition and culture that are evident throughout the continent are the following:

- interconnectedness
- respect for history, ancestors
- collaboration
- informal modes of production
- informal economic structures
- development of local solutions
- inclusion of all members of the community

Table 1.   Three traditional African philosophies.

Philosophy	Application
Ubuntu	
Interconnection; I am because you are	South African Constitution
	South African Economic Empowerment Policies
	Peace and Reconciliation
Harambee	
Pull together; local solutions to local problems, local production	Kenya's post-colonial self-reliant national economic development strategy
Sankofa	
The best way to secure our future is to learn from our past	African-centred leadership in the African diaspora
	Movement toward re-establishing pre-colonial ideas about production, development, and leadership

Although there are many African philosophies and traditions that incorporate the aforementioned themes, for this discussion, three traditional African philosophies have been identified. All three of these philosophies incorporate these themes and additionally have a practical application: *ubuntu*, *harambee*, and *sankofa* (Table 1).

Ubuntu – I am because you are, interconnection (South Africa)

Although the term Ubuntu originates in South African languages, the term can also be found in other African cultures and often represents a view of sharing and collective ownership of opportunities, responsibilities, and challenges, as well as the importance of people and relationships over things.

Harambee – all pull together (Swahili)

Harambee is the official motto of Kenya, and it appears on the country's coat of arms. Kenya's first prime minister and president, Jomo Kenyatta, spoke at the opening of parliament of the newly independent state. He asked the people of Kenya to adopt the spirit of *harambee* and to all pull together to find local solutions to problems. Kenyatta used *harambee* as a call to action, urging people to unite to help build the newly independent nation. Each person is expected to contribute towards the well-being of the clan, according to his or her age, knowledge, skills, and experience.

Sankofa – Draw upon the lessons of the past for present-day solutions (Ghana)

Sankofa is an Akan (Ghanaian) word that means "return to the source and fetch (learn)". In the book *Sankofa: African Thought and Education*, Tedla (1995) urges educators and policymakers in Africa and throughout the diasporas to reach back into the past to rediscover lost traditions. The philosophy of *sankofa* additionally challenges Africans to renew and refine these traditions so that they will have new meaning for all Africans, not just the wealthy and powerful, in both the present and future.

Forms and variations of *ubuntu*, *harambee*, and *sankofa* are evident throughout the continent and have been implemented and revered in various forms, particularly in sub-Saharan Africa. By honouring local, home-grown responses to its unique and often unpredictable contexts, Africa's experiences offer humanity a fresh perspective on the value of interpersonal relationships in an increasingly complex world (M'Rithaa 2008).

Considering the prevalence of the concepts of community, sharing, and collective knowledge and production in some sub-Saharan African traditions, these terms seem to provide a basis for an approach to wicked problems.

Overlapping markers in discourse

An examination of the overlap in the discourse or the similar markers in the literature of the three principles of wiki approaches, wicked problems, and traditional African philosophies may provide insight into innovative approaches to wicked problems in Africa and the world (Figure 1). The terms or phrases that are most commonly found in the literature on these three principals include interconnection, collaboration, collective production, sharing, community, multiple contributors, respect for formal and informal knowledge systems, low or no expense to participate, and local as opposed to universal approaches. Since these terms are consistently found in a discourse analysis of the three principals, it seems that NGOs or other global organisations seeking to address wicked problems should develop programmes which incorporate the principals represented by these terms.

The next challenge is to find examples of organisations that illustrate the application of these intersecting principles. Ushahidi and the Geo-Wiki Project provide a practical model for the incorporation of this model as an approach to solving wicked problems.

Ushahidi – community, innovation, openness

One such innovative social enterprise, Ushahidi, applies a wiki approach while considering traditional African principles to address the wicked problems of voter fraud and election violence in Kenya and globally. Ushahidi, which means "testimony" in Swahili, was founded during the 2007–2008 Kenyan elections. In an effort to address election violence and voter fraud, Ushahidi founders developed an open-source mapping software and solicited a corps of volunteers to report via electronic means (SMS, tweet, web post) any witnessed incidences of voter fraud. Based on this information, Ushahidi was able quantify the occurrences and to work collaboratively with the government to identify the location and the nature of election issues. The Ushahidi platform was

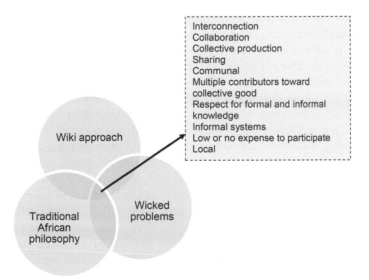

Figure 1. Overlap between the three principles of wiki approaches, wicked problems, and traditional African philosophies may provide innovative approaches to wicked problems.

built as a tool to easily crowd-source information using multiple channels, including SMS, email, Twitter, and the web. This technology has been adapted and used worldwide by activists, news organisations, and everyday citizens.

Ushahidi co-founder and executive director, Julianna Rotich, initially conceived of and currently operates the company according to the tenets of community, innovation, and openness. Rotich (2013 notes that *"we were not experts or humanitarians; we were bloggers faced with difficult times and came together to reflect on both an individual and a communal response"* (Juliana Rotich - Ushahidi Founder, personal communication, October 15, 2013). She reflects on the traditions and values imparted by her Kenyan grandmother, whose ideas of community help her to contextualise her role as an innovator and technology professional. This idea of community led Rotich to recognise that her approach to the work she does and the fact that she always seeks to answer the questions *"What am I making, what am I fixing, and how am I helping others?"* reflect the traditions of the informal economy often present in an African context. Additionally, this approach suggests an inter-dependence with others in the community and a collective responsibility to find solutions.

In the case of Ushahidi, they are incorporating a wiki approach by first assembling a voluntary group with a common interest in addressing election issues in Kenya (collective action). Addition-ally, since global volunteers developed the Ushahidi platform through the use of a global volun-teer network of software developers, Ushahidi demonstrates another component of a wiki approach: peer production. This peer-produced software is open and available globally to govern-ments, which illustrates sharing. By invoking a community-based and communal approach to addressing the wicked problem of election violence and corruption, Ushahidi is demonstrating the value of collective responsibility that is inherent in *ubuntu*. Devising this solution as a local solution that addresses a local issue is specifically related to the ideas of self-reliance rep-resented by *harambee*.

The Geo-Wiki project

Even in this age of satellites and space technology, it is difficult to generate good automated rep-resentations of the Earth's surface. Although satellite imagery allows us to create global maps of land cover – materials such as grass, trees, water, and cities that cover the Earth's surface – at various resolutions from 10 km to 30 m, there are two issues with all the different products that are now currently available. The first is that these products are only between 65 and 75% accurate. Second, when they are compared with one another, there are large spatial disagreements between them (See et al. 2013). Without good baseline information about the Earth's land cover, such as the amount of forest or cropland, how can we possibly predict future needs?

Monitoring food security and evaluating the ability of countries to respond to food shortages requires good baseline information on the spatial distribution of cropland (Foley et al. 2011). One way in which this gap might be filled is to use an alternative to the more conventional top down approach to mapping. Instead of employing automatic and semi- automatic classification algor-ithms, it is possible to use citizens and interested experts to crowd-source this information using a bottom-up approach (See et al. 2013).

According to its website, the Geo-Wiki project is a global network of volunteers who wish to help improve the quality of global land cover maps. Often there are large differences between existing global land cover maps and the actual current ecosystem, which means land-use science lacks crucial accurate data. Volunteers are first asked to review global land cover hotspot maps to determine the potential of additional agricultural land available to grow crops. Volunteers are also asked to review global land cover hotspot map disagreements. Volunteers then file reports, based on what they actually see in Google Earth *and* their local knowledge, to determine if the land cover maps are correct or incorrect. Volunteer input is recorded in a

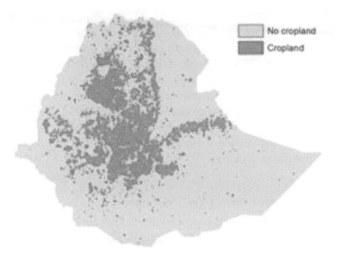

Figure 2.   Map of cropland in Ethiopia created from crowd-sourced data.

database, along with uploaded photos, to be used in the future for the creation of a new and improved hybrid global land cover map. Geo-Wiki founder Steffen Fritz notes that "*this is a task where currently people are still better at than computers – and there is a huge amount of data to look at*" (http://blog.iiasa.ac.at/2014/01/14/interview-taking-geo-wiki-to-the-ground/).

The Geo-Wiki project conducts campaigns and competitions to target specific crop and land cover issues. One such Geo-Wiki campaign focused on creating a cropland map of Ethiopia. The results show that the crowd-sourced cropland map for Ethiopia has a higher overall accuracy than the individual global land cover products for this country (Figure 2). Such an approach has great potential for mapping cropland in other countries where such data do not currently exist. Not only is the approach inexpensive but the data can be collected over a very short period of time using an existing network of volunteers (See et al. 2013). A new project the Geo-Wiki Project is conducting will both employ local experts and train everyday citizens in Kenya to provide on-the-ground information about local land conditions.

The Geo-Wiki Project incorporates a wiki approach by assembling a volunteer group of citizens who are simply using personal computers and in some cases local knowledge to inform land-cover maps (collective action). Additionally, there is low or no cost to participate in Geo-Wiki Project campaigns and contests, since the tools used on the Geo-Wiki site are free and available to anyone interested in assisting with this project. Additionally, by using the expertise of on-the-ground volunteers in Kenya, the Geo-Wiki Project incorporates local wisdom and input to approach the wicked problem of available cropland. The fact that a teacher in the USA, a lawyer in the UK, and a student in Kenya can all contribute to the accuracy of a land cover map illustrates the concept of interconnection, which is a primary principle of *ubuntu*.

Implications of research for practice

There are practical implications of wiki approaches for organisations, businesses, governments, and civil society attempting to address wicked problems. Despite the complexity of wicked problems, considering wiki approaches to development may ultimately diminish the impact of global wicked problems.

Development of local leadership

Malunga (2006) asserts that many development initiatives are imported from the West, and they tend to have only limited success in an African context. As a result, these imported strategies achieve limited success in developing African leaders. He argues that African leadership development should be rooted in African heritage, traditions, and philosophies, especially since African culture is typically ignored or considered an obstacle to good leadership. Since collaboration, community, sharing, and collective production are all integral and authentic components of the African experience, initiatives that include these principles have a better chance to promote ongoing behaviour change. Addressing wicked problems in Africa will require the development of leaders from within African countries.

Promotion of community-based "both ways" education

Additionally, considering local endogenous knowledge allows for community-based "both ways" education experiences. The phrase "both ways learning" indicates the acceptance of a mixing of Western and local indigenous knowledge. Community members become self-oriented participants in the creation of the learning environment, since community-based education begins with people and their immediate reality (Grootjans 1999).

Programme management

As more NGOs and social entrepreneurs seek to manage programmes and organisations that address wicked problems, there will be a greater need for effective management principles to promote success in the field. In order to be effective, management principles and practices should be embraced by their host cultures. Effective management strategies often hinge on the successful harnessing and harmonising of both indigenous and traditional corporate cultures (Mangaliso 2001). Unfortunately, indigenous cultures are frequently suppressed rather than appreciated for their added value. Specifically, traditional management, based on Western corporate culture, has been allowed to dominate management styles around the world at the expense of indigenous cultures. With democracy beginning to dominate around the world, formerly marginalised cultures will increasingly want to express themselves in the workplace. For managers, the challenge is to become familiar with these values and incorporate them into their policies, or they run the risk of out-performance by organisations that do (Mangaliso and Damane 2001).

Democratising development

Wiki approaches open the door for participation that is not hindered by geography or any of the typical barriers that might discourage participation. Often NGOs that are managed remotely do not consider, understand, or effectively utilise local wisdom, which hinders inclusion. Collaborative measures that include sharing, volunteer mobilisation, and collective knowledge place an emphasis on contribution to a solution and pay less attention to the gender, race, and nationality of the volunteers.

Challenges

Although there can be significant advantages to incorporating traditional African philosophies and endogenous wisdom into wiki approaches to wicked problems, there can also be challenges.

- Issues may arise when the locals use traditional philosophies as an excuse or as a basis for resistance to innovation and collaboration.

- "Groupthink" describes instances in which loyalty requires each member in a group to avoid raising controversial issues (Janis 1972, 1982; Rose 2011). Janis (1982, 972) states that groups bring out the worst, as well as the best, in terms of decision-making.
- Engaging in collaborative communities means ceding some control, sharing responsibility, embracing transparency, managing conflict, and accepting that successful projects will take on a life of their own.
- Often, the challenge in scaling up wiki approaches can relate to the amount of information received and the fact that there may be too much information shared collaboratively, rather than not enough.
- Shared knowledge, even indigenous knowledge, must be assessed and evaluated for its relevance.

Conclusions

Both wiki approaches and the application of indigenous wisdom have been considered independently as important factors contributing to global development and solving wicked problems. However, the intersection in discourse between the two approaches provides a way to address complex global problems in a manner that further extends the impact by providing boundary-less and inclusive opportunities for input and production. Further research might include a closer examination of the noted challenges to this approach, including avoiding groupthink and free-loaders. It may be important to understand the issues that arise when traditions and heritage are used as an impediment to innovation.

Acknowledgements

Although not listed as a formal author of this paper, the contribution and research of Camilla Burg was essential in writing this article. Camilla and I conceptualised the "wiki approach" to wicked problems in a café in Paris and continued to collaborate using various wiki tools once I returned to the USA.

Note

1. The book *Tackling Wicked Problems Through the Transdisciplinary Imagination* (Brown, Harris, and Russell 2010) further examines the idea of a transdisciplinary approach to wicked problems. This text approaches inquiry and examination of wicked problems in a way that includes all forms of knowledge that may be relevant to addressing a wicked problem. The edited volume urges practitioners and academics to move toward a new set of innovative solutions using creative imagination as a way to integrate that full range of knowledge into public decision-making. Although this is an important component of any discussion of wicked problems and solutions, it is not the focus of this article. The term transdiciplinary is merely mentioned as a point of consideration for further research and as a justification for suggesting the innovative ideas of this article.

References

Balint, P. J. 2011. *Wicked Environmental Problems: Managing Uncertainty and Conflict*. Washington, DC: Island Press.

Bell, R. H. 2002. *Understanding African Philosophy: A Cross-cultural Approach to Classical and Contemporary Issues*. New York: Routledge.

Benkler, Y. 2003. "Freedom in the Commons." *Duke Law Journal* 52 (6): 1245–1276.

Benkler, Y., and H. Nissenbaum. 2006. "Commons-based Peer Production And Virtue." *Journal of Political Philosophy* 14 (4): 394–419.

Brown, V. A., J. A. Harris, and J. Y. Russell. 2010. *Tackling Wicked Problems through the Transdisciplinary Imagination*. London: Earthscan.

Bryant, A. 2006. "Wiki and the Agora: 'It's Organising, Jim, but Not as We Know It'." *Development in Practice* 16 (6): 559–569.

Coppock, D. L. and S. Desta. 2013. "Collective Action, Innovation, and Wealth Generation Among Settled Pastoral Women in Northern Kenya." *Rangeland Ecology & Management* 66 (1): 91–105.

Foley, J. A. et al. 2011. "Solutions for a Cultivated Planet." *Nature* 478 (7369): 337–42.

Grootaert, C. 2001. "Does Social Capital Help the Poor? A Synthesis of Findings From the Local-level Institutions Studies in Bolivia, Burkina Faso, and Indonesia. World Bank Local-Level Institutions Working Paper No. 10." Washington, DC: World Bank.

Grootjans, J. 1999. *Both Ways and Beyond: In Aboriginal and Torres Strait Islander Health Worker Education*. Sydney: University of Western Sydney, Hawkesbury.

Harris, M., and N. Gorenflo, eds. 2012. *Share or Die*. Gabriola Island, BC: New Society Publishers.

Hitt, W. D. 1998. *The Global Citizen*. Columbus, OH: Battelle Press.

Janis, I. L. 1972. *Victims of Groupthink; A Psychological Study of Foreign-Policy Decisions and Fiascoes*. Boston, MA: Houghton Mifflin.

Janis, I. L. 1982. *Groupthink: Psychological Studies of Policy Decisions and Fiascoes*. 2nd ed. Boston, MA: Houghton Mifflin.

Karlberg, M. R. 2004. *Beyond the Culture of Contest: From Adversarialism to Mutualism in an Age of Interdependence*. Oxford: George Ronald.

Khozein, T., M. Karlberg, and C. Freeman. 2013. "From Competition to Collaboration: Toward a New Framework for Entrepreneurship." In *Creating Good Work: The World's Leading Social Entrepreneurs Show How to Build a Healthy Economy*, edited by R. Schultz, 105–116. New York: Palgrave Macmillan.

Lassiter, J. 2000. "African Culture and Personality: Bad Social Science, Effective Social Activism, or a Call to Reinvent Ethnology?." *African Studies Quarterly* 3 (3): 1–21.

M'Rithaa, M. K. 2008. "Engaging Change: an African Perspective on Designing for Sustainability." Presented at the Changing the Change (CtC) International Conference, Turin, 10–12 July.

Malunga, C. 2006. "Learning Leadership Development From African Cultures: A Personal Perspective." INTRAC Praxis Note 25. Oxford: INTRAC.

Mangaliso, M. P. 2001. "Building Competitive Advantage from 'Ubuntu': Management Lessons from South Africa [and Executive Commentary]." *The Academy of Management Executive* 15 (3): 23–34.

Manguelle, D. E. 2000. "Does Africa Need a Cultural adjustment program?." In *Culture Matters: How Values Shape Human Progress*, edited by S. P. Huntington and L. E. Harrison, 65–77. New York, NY: Basic Books.

Meinzen-Dick, R., and A. Knox. 1999. "Collective Action, Property Rights, and Devolution of Natural Resource Management: A Conceptual Framework." Workshop on Collective Action, Property Rights, and Devolution of Natural Resource, Puerto Azul, Philippines, June.

Mitchell, S., 2008. "Easy Wiki Hosting, Scott Hanselman's blog, and Snagging Screens." *MSDN Magazine*. Accessed 1 October, 2013. http://msdn.microsoft.com/en-us/magazine/cc700339.aspx

Raymond, E. 1999. "The Cathedral and the Bazaar." *Knowledge, Technology & Policy* 12 (3): 23–49.

Raymond, E. 2000. *The Cathedral And The Bazaar*. Sebastapol, CA: O'Reilly.

Rittel, H. W. J., and M. M. Webber. (1973) "Dilemmas in a General Theory of Planning." *Policy Sciences* 4 (2): 155–169.

Rose, J. D. 2011. "Diverse Perspectives on the Groupthink Theory – A Literary Review." *Emerging Leadership Journeys* 4 (1): 37–57.

Schultz, R. 2013. *Creating Good Work: The World's Leading Social Entrepreneurs Show how to Build a Healthy Economy*. New York, NY: Palgrave Macmillan.

See, L., I. McCallum, S. Fritz, C. Perger, F. Kraxner, M. Obersteiner, U. D. Baruah, N. Mili, and N. R. Kalita. 2013. "Mapping Cropland in Ethiopia Using Crowdsourcing." *International Journal of Geosciences* 4 (6): 6–13.

Shirky, C. 2008. *Here Comes Everybody: The Power of Organizing without Organizations*. New York, NY: Penguin Press.

Shutte, A. 1993. *Philosophy for Africa*. Roden bosch: University of Cape Town Press.

Tapscott, D. and A. D. Williams. 2006. *Wikinomics: How Mass Collaboration Changes Everything.* New York, NY: Portfolio.

Tedla, E. 1995. *Sankofa: African thought and Education.* New York, NY: P. Lang.

Vaitheeswaran, V. V. 2012. *Need, Speed, and Greed: How the New Rules of Innovation can Transform Businesses, Propel Nations to Greatness, and Tame the World's Most Wicked Problems.* New York, NY: Harper Business.

Weber, E. P. and A. M. Khademian. 2008. "Wicked Problems, Knowledge Challenges, and Collaborative Capacity Builders in Network Settings." *Public Administration Review* 68 (2): 334–349.

Using Rwandan traditions to strengthen programme and policy implementation

Angélique K. Rwiyereka

Implementing change is far harder than making policy pronouncements that call for change. Rwanda, in the 20 years since the 1994 genocide, has made substantial progress in turning around its economy and in meeting key Millennium Development Goals (MDGs). Real GDP in Rwanda grew at a rate of over 8% per year in the past years, the percentage of the people living in poverty has dropped by 14%, and UNDP reports that Rwanda is on track to meeting many but not all MDGs by 2015. Rwanda's progress in economic and social spheres stands out in Africa, where many countries, despite commitments to the MDGs, lag behind on performance. The difference in Rwanda is the leadership's attention to implementation, and the incorporation of endogenous practices, particularly into planning and accountability. This article is based on observations of practice at national and community levels and of policy design and implementation. It is a by-product of a study of the impact of different approaches to community health delivery systems in Rwanda, completed as part of the author's doctoral dissertation, and also of the author's experience working within the government in Rwanda.

Il est bien plus difficile de mettre en œuvre des changements que de faire des déclarations de politiques générales qui demandent des changements. Le Rwanda, durant les 20 années qui se sont écoulées depuis le génocide de 1994, a accompli des progrès considérables pour ce qui est de rétablir son économie et d'atteindre des Objectifs du Millénaire pour le développement (OMD) clés. Le PIB réel du Rwanda a augmenté de plus de 8 % par an au cours des quelques dernières années, le pourcentage de personnes en situation de pauvreté a diminué de 14 % et le PNUD signale que le Rwanda est en bonne voie pour atteindre de nombreux OMD (mais pas tous) d'ici à 2015. Les progrès réalisés par le Rwanda dans les sphères économiques et sociales se distinguent en Afrique, où de nombreux pays, malgré leur engagement en faveur des OMD, sont à la traîne sur le plan des performances. La différence au Rwanda est l'attention accordée par le leadership à la mise en œuvre, et l'incorporation des pratiques endogènes, en particulier dans la planification et la reddition de comptes (redevabilité). Cet article se base sur les observations des pratiques aux niveaux national et communautaire et de la conception et mise en œuvre des politiques générales. Il s'agit d'un sous-produit d'une étude sur l'impact de différentes approches des systèmes communautaires de prestation de services de santé au Rwanda, effectuée dans le cadre de la thèse doctorale de l'auteur, et également de l'expérience de l'auteur de son travail au sein du gouvernement rwandais.

Resulta mucho más fácil formular declaraciones de políticas que impulsen el cambio que implementarlo. Durante los veinte años posteriores al genocidio ocurrido en Ruanda en 1994, este país ha avanzado sustancialmente en relación a la reactivación de su economía y al cumplimiento de los principales Objetivos de Desarrollo del Milenio (ODM). En los últimos años, el PIB real de Ruanda creció a una tasa anual de más de 8%, mientras que el porcentaje de personas que viven en la pobreza disminuyó en 14%. Asimismo, el PNUD ha

72

informado que el país se encuentra encaminado a alcanzar muchos, mas no todos, los ODM en 2015. Los logros conseguidos por Ruanda en el ámbito socioeconómico resultan ejemplares para África, donde a pesar de haberse comprometido a lograr los ODM, muchos países se encuentran a la zaga en este sentido. En Ruanda, la atención puesta por el liderazgo a la implementación, particularmente a los aspectos de planeación y rendición de cuentas, así como a la incorporación de prácticas endógenas, ha marcado la diferencia. El presente artículo se apoya tanto en la observación de las prácticas establecidas a nivel nacional y comunitario, como en la observación del diseño y la implementación de las políticas. Asimismo, analiza los resultados de un estudio de impacto aplicado a los distintos enfoques desde los cuales se brindan servicios de salud a nivel comunitario, en el contexto de la disertación doctoral de la autora y de las vivencias experimentadas en su trabajo en el ámbito gubernamental de este país.

Rwanda since 1994

Rwanda has demonstrated notable success in growing its economy and improving health status and educational opportunities for its people. In addition to a growth in real GDP of 8% over the years 2001–2012 and a reduction of people living in poverty of 14%, Rwanda is one of the global high performers among very poor countries that have, against substantial odds, made significant progress in meeting social goals (World Bank 2014). Rwanda has achieved a reduction in under-five mortality rates of at least 60% since 1990; 70% of the population are sleeping under insecticide-treated malaria nets; there is universal access to antiretroviral therapy for people with HIV/AIDS (UN 2013). In many areas, it has performed at levels far ahead of its African neighbours, despite the legacies of a genocide and the constraints of being a densely populated, landlocked country. The successes Rwanda has achieved draw in part on leadership commitment and the ability to rely on Rwandan traditions to build a commitment to implementing economic and social policies.

Sources of successful economic and social change

The primary difference between Rwanda and other African countries is the seriousness of top leadership's commitment to making change in the country. In the early years after the genocide, the President worked closely with grassroots groups and through radio, TV, and newspapers; he made sure the development message went out as widely as possible. He also travelled around the country meeting with people wherever they were – in open spaces, in stadiums, or on mountainsides. The message he gave was that Rwanda needed to change; it needed to improve the economic and social conditions of Rwandans; the message was also that culturally, Rwandans did not live by poor standards, and it should not be so now. Rwandans, coming out of the genocide, were largely receptive to this message of change because there was no way to go deeper into poverty. Change was the only option for survival. The President managed to raise people's expectations and to begin to redefine citizens as Rwandans, an image that was tarnished with the genocide. This dialogue across the country allowed Rwandans to build a national commitment to achieving social change and to reaching international and national goals that Rwandans adopted. In this way, MDGs were not just goals set by external agencies, but were merged with goals that Rwandans set for themselves; they are goals Rwandans are committed to meeting and even going beyond. Likewise, the Poverty Reduction Strategies may have been negotiated with outsiders, but they represented Rwandan goals, not goals set solely outside.

The second reason for Rwanda's success has been its ability to *implement* policies, and to hold officials accountable for implementing the policies or goals to which they committed. Rwanda's modern capacity to do this builds on its own history and culture, and the cultural importance of living up to one's commitments. This pattern of national goal-setting and accountability for implementation is rooted in Rwandan traditions. In pre-colonial and even colonial days, Rwanda was a kingdom. A council of knowledgeable elders (and sometimes younger people) called *inyangamugayo* was important; with the rest of the population, they would meet to consider community issues and discuss problems to find solutions, and then make decisions on what would happen. These meetings took place at village level. The *inyangamugayo* were highly respected people because they were known for their high community and personal values. They were making sacred commitments for justice and fairness, for which they would be willing to die to fulfil – people could trust their words because they would die for what they had committed to do. *Inyangamugayo* would be banned from this role if they failed to achieve the highest standard of justice and fairness the community expected from them.

Observers of Rwanda may be aware of the *gacaca* (a word that means grass), which many outsiders understand as a community-based judicial system set up to achieve truth, justice, and reconciliation in the post-genocide period. *Inyangamugayo* were selected by the community to convene the *gacaca*. The *gacaca* would be held in an open, grassy spot to consider an issue, and the *inyangamugayo* would listen and make decisions. Everyone in the community would have the right to speak and proceedings were to be fully transparent. *Gacaca* in current Rwanda were thus more than a means for a court system to work: they were aimed at reuniting and rebuilding communities.

Beyond the *gacaca* court, *inyangamugayo* can still be found in all kinds of societal groups. Commitments in the older kingdom times were made to achieve a goal or goals in time of war or peace, at the level of the community or in small gatherings. Commitments required follow-through and it was because people believed that the *inyangamugayo* would carry out what they promised that they were trusted. Commitments made in this way were called *imihigo* (from the verb *guhiga*), or a set of goals that a person would make in public and then carry out. As defined by the Ministry of Local Government or Minaloc (managers of the *imihigo* programmes), the concept is defined as "*a cultural practice in the ancient tradition of Rwanda where an individual would set himself/herself targets to be achieved within a specific period of time and to do so by following some principles and having determination to overcome the possible challenges*" (Minaloc, 2012).

In 2006, the *imihigo* concept was incorporated in the current Rwandan development agenda; this represents an important engine for change by as it raises expectations and enthusiasm for attaining development goals. From the government perspective, the *imihigo* is an invaluable tool in the planning, accountability, and monitoring and evaluation processes. It is a results-based framework with three pillars: economic development, social development, and governance and justice. It ensures full participation and ownership of citizens. The aim of promoting the *imihigo* is to:

(1) speed up development by implementing national programmes;
(2) promote the culture of providing evidence and showcasing Rwanda's achievements;
(3) promote the culture of working with purpose;
(4) promote the culture of *guhiganwa no guhanga* (competition and innovation);
(5) promote joint planning and coordinated implementation for development;
(6) use all resources and means with the aim of reaching goals as quickly as possible; and
(7) promote the culture of self-evaluation in all endeavours (Government of Rwanda 2011a).

Beyond commitments or *imihigo*, there is another related Rwandan word – *umushikirano* – which means to reach out to someone or to the people (from the verb *gushikirana*). The Government has used this concept to make the end of every year a time to reach out to its population by organising a two- to three-day gathering of leaders from all levels (from the President to the head of the village), to respond to the population's concerns and questions. It is in this forum that goals or *imihigo*, and the leaders carrying the responsibility to reach these goals, are assessed for the level of success. During this live event, the conversations between the leaders and the population are shared on TV, on radio, online, and using telephone short messaging systems. Every leader is accountable directly to the population in this process. Topics raised may include questions about the community-based health insurance, access to loans, performance contracts, or use of education funds, among others.

Traditional concepts and incentives for high performance

The Government in Rwanda does use traditional practices to create incentives at all levels of decision-making not just to set goals but to implement them. The incentives are not like those in Western management practice, which are often financial. In Rwanda, *imihigo* is a method of setting goals and evaluating performance results, but it works differently because it is embedded in Rwandan traditions. Traditionally the people, king, and chiefs would make public statements of what they would deliver to the people, and they were expected to follow through on their words. Today, leaders commit to being *inyangamugayo* and at different levels they participate annually in setting goals or *imihigo*. What drives them to meet the goals is not a financial reward, but a fear of failing to achieve the goals, because it would be shameful to fail to live up to promises made – it would mean one had failed to live up to the trust placed in him or her, and trust was traditionally key to being *inyangamugayo*. It is not so much the fear of losing one's position but the fear of loss of *reputation*, and that one's failure to deliver on promises will follow one around for the rest of one's life. *Imihigo* is one part of a larger system that focuses on performance results and accountability at district level.

At the central level, Ministers participate in setting, and are bound by, national goals aimed at reaching the Economic Development and Poverty Reduction Strategies (EDPRS) and the Millennium Development Goals (MDGs) – they are also bound by the national *imihigo*. At this level, the *imihigo* are evaluated through *umwiherero*, which is a central Government retreat that covers three days every year with the aim of discussing achievements, challenges, and new targets. Unlike the district-level *umushikirano*, this is a closed meeting among the President and Ministers, Ministers of States, and relevant officials such as Permanent Secretaries, Rectors of national universities, and others. The *umwiherero* is held far from the city with no access to media, so that top leaders are free to evaluate national annual goals and their implementation, to review what went wrong and what went right, and to speak honestly with each other. The top leaders also set new programmes and new targets, and track their progress towards achieving the country's vision 2020. A summary of this planned way forward is published through media. Ministry leaders, having participated in this national retreat, also take the national goal-setting to practical levels.

The Ministry of Health, for example, sets specific goals, merging local goals with those set in international commitments like the MDGs; priority may be given to assuring that pregnant women who are HIV-positive have access to measures to prevent transmission of HIV from mother to child, for instance. The Ministry of Finance consolidates goals and matches them with funding available locally and from donors. It is this internal processing of goal-setting and evaluation that gives the Ministries the ability to negotiate with multiple donors who may have conflicting requirements. An example of coordinated leadership in health is the procurement of antiretrovirals (ARTs). Rwanda receives funding from the Global Fund to Fight HIV, Tuberculosis, and Malaria; the United States President's Emergency Plan for AIDS Relief, and others. Each donor may, for example, have their own ideas on how to procure ARTs and how to deliver services, but the Rwandan Government was able to argue successfully with donors that having a common procurement mechanism and services integrated at health centre level was the best way to go. Purchasing ARTs from a single, lowest-cost source, was key to the sustainability of these programmes after donor assistance declined. That is how the common basket was created, so that concerned donors in HIV care and treatment would pool resources and purchase ART together.

Based on the national goals set during the *umwihererano*, there are also meetings at the local level to set goals or *imihigo* for the coming year, both at district and village level. The evaluation of *imihigo* brings in rewards at district level; districts are ranked by performance and the best performers are rewarded, because it is important to keep momentum toward the goals. In this process, district mayors compete for achieving agreed-on targets on specific, measureable indicators. Leaders do not want to fail to live up to their commitments. As the *umushikirano* is a live event and *umwihererano* is an open frank discussion on policy, programmes, and implementation, it is not possible to know what issues will be brought up by the people and president, and one needs to be prepared to show that issues raised have been handled or will be considered adequately. This creates an incentive for the officials to pay attention to assuring that issues raised are adequately addressed.

There is another concept, *umuganda*, which brings national *imihigo* to the household level. *Umuganda* is community work, where traditionally people would gather as a group to provide free labour for the vulnerable members of the community. Every last Saturday of the month, people gather in their villages and work in community; after work, neighbours – including ministers and leaders at all levels – sit and discuss national goals, issues, and possible solutions and apply them to their local context. This allows rapid and effective communication between central and local levels because it takes only a few hours for Ministers to communicate with District mayors, and mayors to send the message to sectors' executive secretaries, from there to the cell and village levels. In the case of HIV/AIDs goals, for example, leaders at lower levels may commit to using resources (both human and financial) to reach out to the underserved AIDS and non-AIDS Health Centres. In this way, Rwandan traditions of honouring commitments to goals are merged with effective use of HIV funds and performance-based financing techniques. This integration of modern and traditional practices has helped, in this case, to reinforce the reach of non-AIDS services and non-AIDS Health Centres through outreach programmes (Shepard et al. 2012). The whole process of goal-setting, implementation, and evaluation can be described as a development cycle based on traditions, as illustrated by the chart in Figure 1.

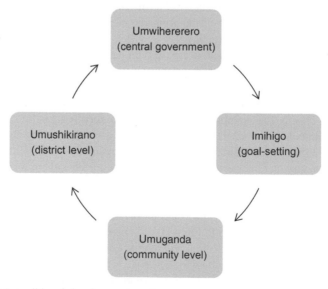

Figure 1.   Rwanda's traditional development cycle.

Challenges in making *imihigo* work

Everything in Rwanda has been more or less experimental, because Rwanda used original recipes to spearhead its own development. Because Rwandans are humans, we expect to find mistakes in the institutions being developed. It is not easy to draw on traditional culture to shape modern management practices, and traditional cultural practices alone are not enough to assure achievement of social goals. It is necessary to take full advantage of scientific approaches to designing and implementing service delivery, but incorporating traditional practices has important benefits for Rwanda. Having a shared discussion of goals and participation in setting goals allows the government to develop performance contracts with government officials that carry the force of a traditional commitment (*imihigo*). When there is a public, transparent system for evaluating performance (*umushikirano*), public officials can be held accountable for performance in an open way. As government officials demonstrate that they are accountable and delivering on the performance they promised, a trust can be developed between government and the people. These characteristics of openness, participation, and most importantly trust, are critical to building a sustainable society in Rwanda.

Conclusion

An evaluation report of 2011 showed that there were many improvements to be made to the imihigo system; however, *"on the whole, the concept of Imigiho as a development strategy has led to promising results by promoting a competitive spirit and creating focused and enthusiastic effort which are essential ingredients to sustainable development"* (Government of Rwanda 2011b, 15). Wu Zeng, at the Heller School at Brandeis University, analysed several databases from Brandeis and UNAIDS and found that governance – an aspect of imihigo – was the most important element in influencing efficiency in health service delivery, followed by financing mechanisms (Wu 2009). My own practical observations about Rwanda's governance system, accountability, and the role of traditional participatory and accountability practices correspond to Wu's findings. My own dissertation found that mayors' performance-based contract was a driving force behind the performance of Rwanda in using available money from donors and the government to improve maternal health indicators (Rwiyereka, A.K., 2013). Because people respond intrinsically to their own culture, talking about development in their indigenous language and using indigenous concepts inspires them more than the use of foreign words or attempts to internalise foreign concepts. Rwandans relate to their development programmes because these refer to their culture; they cannot give an excuse of not honouring their culture or failing to achieve what their elders managed to do. There is this drive everywhere, but strongly in Africa, that people follow the steps of elders, or they are judged if they fail to make happen what elders started – it is a continuity instinct. In brief, it is easy and effective to talk development in local language, based on local culture.

While what Rwanda is attempting needs to be refined and improved, there are some lessons that can be shared with other African countries. The lesson Rwanda imparts from using traditions to speed development suggests that if Africa had been allowed time to develop naturally, without the blank slate approach of the colonialist hand (which assumed Africans had nothing positive to share as a foundation for progress), then locally-grown, positive development approaches could have naturally emerged. The last century, because of slavery and colonialism, eroded some traditions, but it is not too late for Africa to build on its cultures and traditions to move forward and become stronger. That is why it is critical to document culture, traditions, and beliefs as the elders who know our history are passing away.

References

Government of Rwanda. 2011a. *National Policy Document on Imihigo*. Kigali: Government of Rwanda.

Government of Rwanda. 2011b. *Local Government Imihigo Evaluation Report 2010–2011*. Kigali: Government of Rwanda.

Minaloc. 2012. *Imihigo Concept Note*. Kigali: Ministry of Local Government.

Rwiyereka, A. K. 2013. "Making Money Work for Mothers: A Quantitative and Qualitative Assessment of the Impact of Novel Health Financing Policies on Maternal Health Services in Rwanda". PhD diss., The Heller School for Social Policy and Management, Brandeis University.

Shepard, D. S., Z. Wu, P. Amico, A. K. Rwiyereka, and C. Avila-Figueroa. 2012. "A Controlled Study of Funding for Human Immunodeficiency Virus/Acquired Immunodeficiency Syndrome as Resource Capacity Building in the Health System in Rwanda." *American Journal of Tropical Medicine and Hygiene* 86 (5): 902–907.

UN. 2013. *The Millennium Development Goals Report 2013*. New York: the United Nations.

UNDP. n.d. "Rwanda Millennium Development Goals". United Nations Development Programme. Accessed December 16, 2013. http://www.unrwanda.org/undp/mdg.htm

World Bank. 2014. "Rwanda Overview". Accessed December 6, 2013. http://www.worldbank.org/en/country/rwanda/overview

Wu, Z. 2009. "Resource Needs and Performance of National HIV/AIDS Programs in Low- and Middle-Income Countries." PhD diss., The Heller School for Social Policy and Management, Brandeis University.

Lessons of endogenous leadership in Nigeria: innovating to reduce waste and raise incomes in the cassava processing and goat-keeping systems

Danielle Fuller-Wimbush and Kolawole Adebayo

Agricultural innovations are increasingly emerging from African research scientists. This paper looks at an innovation developed by animal scientists at the University of Agriculture Abeokuta in Nigeria. They identified a method for drying cassava peels, which creates an income source for rural women, reduces environmental waste, and raises the income of goat herders by transforming the cassava waste into animal feed. Initially funded by a World Bank grant, this paper addresses the challenges of securing donor funding for local innovations and presents an argument for a new model of development that supports locally driven solutions to current development issues.

Des innovations agricoles émanent de plus en plus des chercheurs scientifiques africains. Cet article traite d'une innovation mise au point par des zoologues de l'University of Agriculture Abeokuta du Nigéria. Ils ont identifié une méthode pour faire sécher les épluchures de manioc qui crée une source de revenus pour les femmes en milieu rural, réduit la quantité de déchets déversés dans l'environnement et accroît les revenus des éleveurs de chèvres en transformant les déchets du manioc en aliments pour leurs bêtes. Cette initiative a été financée dans un premier temps par une subvention de la Banque mondiale, et cet article traite du défi d'obtenir des fonds auprès des bailleurs de fonds pour des innovations locales et présente un argument pour un nouveau modèle de développement qui soutient des solutions impulsées au niveau local à des problèmes de développement actuels.

Los científicos investigadores de África producen cada vez más innovaciones en el área agrícola. El presente artículo examina una de las innovaciones desarrolladas por científicos de la Universidad de la Agricultura en Abeokuta, Nigeria, especializados en animales. Estos elaboraron un método para deshidratar cáscaras de yuca, que permite transformar los desperdicios de yuca en alimentos para animales. A la vez que este método crea una fuente de ingresos para las campesinas, se reducen los desperdicios ambientales y se eleva el ingreso obtenido por los pastores de cabras. El presente artículo, financiado inicialmente por el Banco Mundial, examina, por un lado, los retos que implica asegurar el financiamiento de donantes para destinarlo al apoyo de las innovaciones locales y, por otro, presenta argumentos orientados a promover un nuevo modelo de desarrollo tendiente a apoyar soluciones locales a los actuales problemas presentes en el ámbito de desarrollo.

Introduction

As the largest producer of cassava in the world, Nigeria is faced with the challenge of how to properly dispose of the high volume of residual waste that is created when turning the root into flour and other edible products. Often, the cassava peels are burned or left to rot, damaging the environment and polluting the air. A solution was developed and tested by a team of Nigerian animal scientists and rural development experts from the University of Agriculture Abeokuta, in collaboration with government extension workers, that utilises the discarded cassava peels in a way that reduces waste and at the same time generates income for goat herders and resource-poor women working in the cassava processing centres.

The animal scientists recognised that the nutrient-dense cassava peels, if properly dried, could be used for animal feed. Utilising this knowledge, they developed a special diet for goats that maintains the health of the animals, improves the growth rate, and reduces the cost of antibiotics and risk of death. The nutrient-dense diet, comprised of 30% dried cassava peels and 70% grasses, legumes, and roughage, increases the growth rate, reducing the time it takes to bring the goats to market size by half, and cutting the typical cost of feed by close to US$198 a year (Development Marketplace Proposal 2008).

Case study: innovation in Nigerian cassava industry and goat keeping systems

In order to transform the discarded cassava peels into dried feed, the local scientists designed a drying platform for the cassava peels that would allow the discarded waste to dry quickly in the sun, and be removed from the ground, keeping it clean of dirt and other debris. Once dried, the cassava peels can be packaged and sold in the market to goat farmers who can then use the peels to feed their goats. The drying platforms are made from a cement base with a black tarp finish, which absorbs the sun, and are able to be constructed by local builders in the rural communities. Local women working in the cassava mills are trained to dry the cassava and sell it in the market to generate additional income. The farmers benefit by having readily available cassava to add to their animal feed, without having to gather and dry the cassava themselves. When included in the animal feed, the goats are healthier and grow faster, reducing the amount of time required for the animals to reach maturity and be sold in the market. This results in an economic gain for the goat farmer who also spends less in antibiotics and feed for the animal.

Kolawole Adebayo, a trained rural development expert from the University of Agriculture Abeokuta, saw the potential to expand this innovation, creating a win-win proposition: reduced environmental waste, increased income for resource-poor women, and financial savings for goat farmers. Because the model was profit-making, it could eventually be sustainable but would require initial seed money to get started. With his initiative, the University succeeded in receiving a grant from the World Bank, as part of their Development Marketplace initiative to provide small grants (under US$200,000) to fund pilot projects in developing countries. The funding allowed the team of Nigerian scientists and researchers to develop a five-part innovation that could be scaled up throughout the country. The programme includes a *technology*, in the form of a drying platform for the cassava peels; a *new product*, clean dried cassava peels packaged and sold as goat feed; an *educational component*, in the form of training for the goat farmers on the cassava based diet that maximises the growth rate and health of the animals; *access to credit*, an essential component for resource-poor farmers or cassava processors to be able to build the drying platforms; and a new *market mechanism* to link cassava processors and goat keepers.

The pilot was successful, demonstrating a measurable economic benefit to both cassava processors and goat farmers with average incomes of less than US$2 a day. Initial findings showed gains in income of US$384 a year for cassava processors and US$198 a year for goat farmers (Development Marketplace Team 2010).[1] Scaling up these gains throughout the cassava growing regions of Nigeria can help the country meet its poverty reduction and improved livelihood goals, at the same time that it reduces the impact of burning cassava waste on the environment.

Demand for participation in the cassava-drying project is strong, with farmers and cassava processors recognising the economic benefits that are possible. In its first two years, the project exceeded the initial target of reaching 3,600 cassava processors and 600 goat keepers, expanding its reach to 21 additional locations and directly benefiting 6,078 processors and 886 goat keepers. The expansion was a result of pressure from agricultural extension officers and the communities they represent. Project team leaders estimated that there is potential to reach 200,000 of the 350,000 farmers in the state of Ogun, in addition to thousands more in other states within the country (Development Marketplace Proposal 2008).

From a development perspective, the project produces social change by increasing the income of poor rural farmers, providing them with the resources for basic necessities such as food, health care, and school supplies, improving their quality of life and reducing the burden of poverty. The majority of cassava dryers are women who use the additional income to cover basic needs for their families. Since the project targets vulnerable populations living at the margins of poverty, the additional income is able to improve the living conditions of thousands of families. Likewise, the goat farmers also benefit by raising healthier goats in half the time, providing additional annual income that is often used to expand their business and purchase additional goats.

Bringing the project to scale[2]

The challenge in many developing countries like Nigeria is not a lack of local innovation, but a lack of access to capital to develop and expand the innovation. In the case of the cassava-drying project, the model has been tested and has been shown to have the capacity to scale throughout rural communities within Nigeria and beyond.

More than 228 million tons of cassava was produced worldwide in 2007, with 52% of world-wide production originating from Africa. As the largest producer in the world, Nigeria alone accounts for 46 million tons annually (IITA 2013). The opportunity to scale the innovation is vast but requires additional funding to expand to new areas. As a recipient of World Bank funds, the project was tested and proven viable. A 2010 assessment from the World Bank showed that the project was ready to scale but would require local leadership and financial assistance (Fuller 2011). Like many worthwhile innovations, the challenge lies in securing the requisite funding to continue and grow. The opportunity for donors is to capitalise on local innovations that have demonstrated local support and viability.

In the case of the cassava project, the work has been able to continue beyond the initial pilot phase, partly through support from another World Bank project in Nigeria: the Third National Fadama Development Project (popularly called FADAMA III) and partly through the European Union-funded research project: Gains from Losses of Root and Tuber Crops (popularly called the GRATITUDE Project).

The Third National Fadama Development Project (FADAMA III) is funded through a credit of US$250 million received from the International Development Agency (IDA). The project is anchored in the community-driven development (CDD) approach, which creates sound conditions for sustainability of the smallholder development strategy. It tries to stimulate broad-based growth and expand employment opportunities in the non-oil economy. After the successful pilot testing and evaluation of the cassava waste project, the World Bank's Task Team Leader in Nigeria for the FADAMA Project convened a zonal meeting of the project to review and adapt the interventions proposed to other States of Nigeria. As a result, through mainly lateral diffusion among agricultural extension, the project's intervention is now benefiting poor cassava processors and goat farmers in 10 additional states of Nigeria. This arrangement does not bring additional resources to the project implementers, but it facilitates greater awareness of project interventions and makes the process of natural diffusion faster than it would have been otherwise.

The second source of funding was provided by the GRATITUDE Project, a collaborative research project of the Natural Resources Institute (NRI) of the University of Greenwich and 16 collaborating and implementing institutions from six countries in Europe, Africa, and Asia, inaugurated in 2012 with a grant of US$3.6 million.[3] The project targets cassava and yam as important food security crops for approximately 700 million people around the world who often experience significant post-harvest losses. The project aims to reduce these losses in order to enhance the role that the crops play in food and income security.

The University of Agriculture Abeokuta was assigned a portion of the grant to develop and validate methods for turning cassava waste materials into animal feed suitable for goats. This includes the pre-treatment of waste with fungi to increase the digestibility. Preliminary results from this research have shown that dried cassava peels are a viable product that farmers are willing to pay for as animal feed. The average quantity of cassava peels at processing centres in Osun and Ogun States were estimated to be 263 kg/day and 247 kg/day respectively, showing a sufficient source of base material to utilise for feed. The research also found that farmers' age, years of education, marital status, and farming system are factors that influence their willingness to pay for the use of fortified cassava peel as feed for goats. The study found that the amount that farmers are willing to pay is US$1/25 kg of dried cassava and US$7/25 kg of fortified cassava peel. The study shows an additional income of US$0.33/day for cassava processors selling only dried cassava peels and an additional US$1.75/day for fortified dried cassava peels.

Challenge of securing donor support

Even though both the FADAMA III and the GRATITUDE projects provided some means of keeping the interventions of the cassava waste project in the public domain, they did not provide the type of funding that could make a successfully piloted local innovation with measurable economic benefit, reach out to the large number of potential adopters. There are many challenges that impact the ability to secure funding for local innovativeness. This includes the non-familiarity of the international community with local innovators, which makes them sceptical of their innovations. Conversely, many local innovators cannot "speak the language" of the international community, so packaging their ideas in ways that would attract international funding could present a challenge.[4]

Also, international funders often demand evidence of management of past projects of a similar magnitude. In most cases, local innovators have never had such an opportunity, so the cycle continues; no experience in managing large international funds, no opportunity to gain such experience, and therefore no funding. Competitions like the Development Marketplace have tried to provide the platform for local innovators to present and pilot their ideas, but the support to take the ideas beyond the pilot stage is still a challenge that needs to be addressed.

Another key problem is that donors use a project approach, which means that when the project ends there is no longer any money, but also no continuing involvement of the donor agency. What may be needed is transition support, where donors and other stakeholders can use their influence to get innovations incorporated into government practice and/or donor practice and help secure additional investors to bring the innovation to scale.

Finally, the problem of national insecurity, particularly in developing countries, may distract the attention of the international community from the innovative developmental work going on within the country. The international community, influenced by reports of war, political instability and other problems, may discourage or prohibit staff of international organisations from travelling to regions experiencing conflict or even to some countries entirely. Where such restrictions are in force, donor staff are unable to establish relationships with local institutions and potential innovators, making it difficult for endogenous innovation to secure funding and support.

Addressing setbacks

Beyond the challenge of securing international funding, innovations often experience setbacks, like any start-up, during which they could benefit from outside partnerships and expertise. In the case of the cassava-drying project, one challenge that plagued its implementation was the management of microcredit. While microcredit is available in many parts of Nigeria, the standard interest rate is 22% and the loans require collateral. This model restricts the loans to wealthier members of the community with greater resources to use for collateral and repayment, excluding the most vulnerable members of the population that it is meant to benefit. This project was initially able to partner with an NGO (SLIDEN AFRICA) that provides microcredit loans at a one-time fee of 10% and without the burdensome restrictions of traditional loans. Unfortunately, without proper accountability measures built in, the project experienced a high default rate of 68% for platform construction loans and 35% for livelihood loans. The credit programme was subsequently put on hold.

Other models of microcredit lending that rely on social accountability mechanisms, such as lending to members of a group or association and combining the loans with financial education classes, have been successful in other communities at maintaining low interest rates and could be explored further for application with this project. Many local innovations fail due to the setbacks that invariably arise. Donors, therefore have a role, not only in providing funding, but also working with the project leadership to facilitate partnerships and learning to overcome potential obstacles, and to allow for these kinds of challenges without cutting funding altogether. Innovation is not bred in the security of certainty but brings a certain degree of risk that donors have to be willing to withstand in order to see results.

Conclusion

Progress requires new ideas and ways of doing things. In the development field, we have an untapped resource for ideas and innovations which we can allow to grow, flourish, and solve local challenges. Instead of looking to the tried and tested models from the West, the development model of the future has the opportunity to invest in local knowledge and innovation, ensuring greater sustainability, where solutions reflect the local context and have widespread support, and where local insight is respected and developed, breeding a culture of innovation and self-sufficiency. Support can come in many forms: as funding for pilot projects to test the viability of local innovation, as long-term sustained support to grow and scale up the innovation to new markets, and the knowledge base and network to garner partnerships when needed. As evidenced by the cassava drying project in Nigeria, communities often have innovative solutions that can be low cost with high yields, reflecting the local understanding of the issue and context. What is needed from the donor community to sustain the implementation and adoption process, is to support communities in addressing their own needs.

Acknowledgements

The authors would like to acknowledge Dr Susan Holcombe (Brandeis University) for her scholarship on scaling up local innovations, and her editorial feedback that contributed to this piece.

Notes

1. This practice note only looks at the innovation period up through 2010.
2. A literature review, to which the first author contributed, is available in the final World Bank publications on scaling up Development Marketplace projects: "Lessons from Practice: Assessing Scalability", available at http://documents.worldbank.org/curated/en/2012/05/16632707/lessons-practice-assessing-scalability
3. The grant was for €2.8 million, equal to approximately US$3.6 million.
4. For discussion of the variables needed to successfully scale a project, refer to the World Bank publication "Lessons from Practice: Assessing Scalability."

References

Fuller, D. 2011. *Case Study on Potential for Scaling Up: Adding Value to Waste in the Cassava Processing-Goat Keeping Systems in Nigeria.* Washington, DC: the World Bank. Accessed May 15, 2014. http://siteresources.worldbank.org/INTARD/Resources/335807-1338987609349/NigeriaCaseStudy.pdf

International Institute of Tropical Agriculture (IITA). "Cassava." Accessed June 20, 2013. http://www.iita.org/cassava

World Bank Development Marketplace Proposal #4345. (2008). Nigeria: University of Agriculture, Abeokuta (UNAAB). Accessed May 14, 2014. http://wbi.worldbank.org/developmentmarketplace/idea/using-cassava-waste-raise-goats

World Bank Development Marketplace Team Year 1 Progress Report (October 1, 2010).

Centring African culture in water, sanitation, and hygiene development praxis in Ghana: a case for endogenous development

Afia S. Zakiya

International development aid is driven by actors steeped in Western neo-liberal theory and practice. Africa has largely received failed Western aid, administered mainly through international NGOs in neo-comprador relationships. This article calls for African-centred and -led development, revitalised through endogenous development (ED) praxis. Using a water, sanitation, and hygiene (WASH) sector case study from Ghana, the article theorises Africa's WASH development within the context of globalisation and the politics of knowledge production on Africa. It shows how ED provides African people with self-determining and culturally relevant development necessary for WASH justice and improved health and livelihoods.

L'aide internationale au développement est impulsée par des acteurs pétris des théories et pratiques néolibérales occidentales. L'Afrique a principalement reçu une aide inefficace de la part de l'Occident, largement administrée par des ONG internationales dans le cadre de relations « néo-comprador ». Cet article lance un appel au développement centré sur l'Afrique et impulsée par elle, revitalisé à travers une praxie de développement endogène (DE). L'auteur se sert d'une étude de cas du secteur AEPHA (approvisionnement en eau potable, hygiène et assainissement, WASH en anglais) du Ghana pour théoriser sur le développement du secteur AEPHA en Afrique dans le contexte de la mondialisation et de la politique de production de connaissances sur l'Afrique. Elle montre comment le DE confère aux Africains le développement autodéterminant et culturellement pertinent nécessaire pour garantir la justice dans le secteur AEPHA et l'amélioration de la santé et des moyens de subsistance.

A nivel internacional, la ayuda orientada al desarrollo es impulsada por actores inmersos en la teoría y en la práctica neoliberales de Occidente. En general, África ha recibido ayuda fallida de Occidente, la cual ha sido administrada por ONG internacionales a partir de relaciones tipo *neocomprador*. El presente artículo sostiene que el desarrollo deberá centrarse en el desarrollo impulsado propiamente desde África, a través de una revitalizada práctica de desarrollo endógeno (DE). Apoyándose en un estudio de caso implementado a nivel del sector de agua, saneamiento e higiene (WASH) en Ghana, el artículo examina la teoría que respalda el desarrollo de este sector en el contexto de la globalización y de las prácticas políticas vinculadas a la creación de conocimientos sobre África. Da cuenta de la manera en que el DE brinda un desarrollo autodeterminado y culturalmente relevante a los pueblos africanos, el cual se considera necesario tanto para lograr la justicia en el sector WASH, como para mejorar la salud y los medios de vida de la población.

Introduction

The proliferation of NGOs[1] since the 1980s "NGO decade" continues today, undeterred by the current global economic crisis and questioning of aid effectiveness. This growth of NGOs is happening in Western and European nations as well as on the African continent. Three critical issues are examined here regarding this global phenomenon, to draw out related lessons for development practice in the water, sanitation, and hygiene (WASH) sector, with an accent on the Ghanaian context. The first issue is an old debate on whose ideology, culture, and knowledge should shape how international NGOs (INGOs/NNGOs), local African NGOs (ANGOs), governments, and the donors that fund their activities understand the core problems and challenges of Africa's development. A second issue is whether or not NGOs, or at least significant numbers of NGOs, especially in Africa, have shed their role as "comprador" institutions and instead respond to African people's needs.

Following from these, the final point discussed is the promising alternative epistemological approach to Western neo-liberal development that endogenous development (ED) represents. This article examines it to determine its value in creating sustainable, culturally-relevant solutions to WASH issues in Africa, in this case in Ghana. As will be shown, there is gross inequality in access, provision, and use of WASH services on the continent, leading to serious health challenges. One purpose of this article is to explore the context of an evolving change effort to facilitate using an ED approach in an international and local NGO's practice, and in the wider WASH development sector. After exploring the development trends and defining critical concepts, we examine a case study of programmatic and institutional change efforts to introduce ED, including preliminary successes and challenges to date. However, a major purpose of this discussion is to draw attention to a dilemma yet to be resolved: the perpetuation of status quo development by INGOs and African NGOs, *despite* the alternative paradigm and approach to development that ED represents. Awareness of this is the first step towards resolving the contradictions of how the political economy of sustainable development praxis is globally rooted in the interests of structural capitalism, quite apart from the interests of the local communities it is nominally helping. The concluding section summarises the core issues discussed and provides some recommendations for future ED scholars and practitioners to consider.

Theoretical and concepts issues: a review of the literature

A selected review of the literature on INGOs, local NGOs, foundations, and various development actors' roles in development praxis is important. The concept of "development" as Aubrey (1997) has shown, is highly contested and fluid. Her work shows how INGOs and their "partners" in Africa, and indeed worldwide, mainly follow Western definitions and approaches to human development, without questioning the theories of change, epistemologies, or processes used that have shown to historically and primarily serve global capitalist interests.

Defining development

Despite proclaiming the "death of development", the era of globalisation has come into being "*with hopes of increased wealth everywhere, providing fresh oxygen for the floundering development creed*" Sachs (2010, 8). Indeed, despite the relevant historical critiques of development theories and practice as being imperialist and biased towards Western interests (Rodney 1971), most contemporary scholars and practitioners see development as both a concept and process that involves the partnership and cooperation of groups, individuals, states, and other institutions, particularly NGOs, all acting in concert to promote development free of ideological baggage. However, the concept of development as initially conceived in the West still remains steeped in neoliberal culture and embedded in an ideology of unidirectional progress: from the purportedly more advanced civilisations of the West/North/Europe, whose science, politics, and economic systems are considered superior, to supposedly backward and inferior peoples in the "South".

It is clear that "development" is a value-laden term that sets out *"who gets, what, when, how"*, how much, under what conditions, and at what costs (Aubrey 1997). It is a political ideology and process that ultimately shapes power dynamics and the global socio-economic order. INGOs and their local partners involved in development projects participate in the political economy of knowledge production from conceptualisation to implementation and evaluation that continuously provides the *"fresh oxygen"* Sachs (2010, viii) mentions that keeps the development sector alive and neo-liberal development paradigms resilient. This all contributes to unequal global relations and control of knowledge production. The constant creation of literature, stories, and research that claims success from Western science and development approaches remains at the core of what keeps the development industry alive. Today, the rise of philanthro-capitalism brings a new form of legitimacy to the Western elitist development agenda, and justifies foreign interventions in Africa.

Development financing practice today is linked with Western concerns about the prevention of global terrorism. The redistribution of risk, rather than the redistribution of wealth, now dominates the international agenda, with Africa remaining an area of focus to prevent Western loss of power (Sachs 2010, 18). The recent merger of Western and European aid agencies with departments of foreign affairs and trade, as with Canada and Australia, is one such example, which merits further investigation.

INGOs as neo-compradors in development

A wide range of literature has emerged over the past quarter century concerning whether Western/European NGOs are continuing a process of neo-colonialism, dominance, and hegemony through NGOs in the South, including African NGOs, that act as neo-comprador institutions. The works of Aubrey (1997), Hearn (2001, 2007), Mohan (2002), Fernando (2003), and others, along with my own personal experiences in the INGO world in multiple contexts, provide the next topic to explore.

From the early 1920s to 1980s, Marxists, dependency theorists, and radical scholars have all conceptualised the term *comprador* beyond its literal Portuguese meaning of "buyer" or "agent" to describe local individuals part of, or aspiring to be part of, an elite or petty bourgeois class working in service of capitalism and the former colonisers of their countries. Compradors uphold imperialism in exchange for resources, positions, and delegated power. Today, Africa is experiencing high levels of Northern intervention unseen since colonialism. The 1980's NGO decade explosion, while purporting to bring solutions to development challenges of the South, actually happened simultaneously with a new era of indebtedness, poverty, aid dependency, and policy conditionality. As NGOs grew exponentially, they helped to fundamentally transform the balance of power between the North and Africa, with Africa the loser. Fowler (2000) has explored the proliferation and role of I/NGOs, while Mohan (2002) and Hearn's (2007) similar works focused on Ghanaian NGOs in particular. Their works examine "civil society" as conceived by Western international NGOs, political scientist/scholars, and practitioners, among others, and has shaped contemporary debates around whether or not INGOs and/or African/local NGOs are sites of resistance, working on behalf of the poor and marginalised, or are instead comprador institutions complicit in maintaining Western global dominance.

Comprador NGOs *"ignore the reciprocal linkages between state and society, the constraining effects of market forces, and the underlying ideological agenda of the major lenders"* (Mohan

2002, 2). A fundamental theory of change embedded in the "partnerships" between INGOs and ANGOs is that ANGOs understand the culture, language, and needs of the communities they work in and have positive relationships with; they can thus facilitate or embed Western ideas, practices, and desired social change (on behalf of INGOs). This belief is debatable, and ignores pre-existing indigenous civil society structures at the local or rural level (e.g., a*safo* groups, *susu* groups, clan networks, women's groups), through which poor rural families organise their social, spiritual, economic, and political lives. After examining this process in Ghana, Mohan (2002) concluded that NGOs in Ghana primarily serve as comprador institutions, acting in the interests of the West (and themselves), in selfish ways and often at the expense of communities they claim to represent. Clearly, the notion of an independent civil society is debatable, and remains so in this era of shifts in philanthropic giving.

Fernando (2003) also explores NGOs and neo-compradorism. He warns of the real struggle to keep ED and indigenous knowledge (IK) from being just another development fad to be accommodated while still serving the interests of Western neoliberal capitalism. Referring to the contradictions of knowledge production by NGOs of IK for sustainable development, he states:

"For those who still function within the mainstream paradigms of development, IK is an important source of overcoming the limitations and failures of development. For the adherents of the post development school, IK is an important source of imagining and realizing alternative paradigms of development. For the culturalists, IK is an invaluable source of politics of difference/politics of identity. For multinational corporations concerned with sustainable development, IK is a highly promising area of investment, a way of making their investments socially responsible. IK has evolved as a means of legitimizing as well as resisting mainstream development. The political power of IK lies in its amenability to such different projects – all claiming their aim is to work toward sustainable development." (Fernando 2003, 55)

What is to be done so that ED/IK does not become co-opted by neo-compradors, and become irrelevant to the very people it should serve? There is a clear need for peoples of Africa to engage in knowledge production and sharing on what they think is a more relevant pathway to development, despite the desire of some to promote the belief that globalisation is invincible, and that Western powers can access all corners of the earth, appropriating the IK of people with impunity.

Philanthrocapitalism

Private foundations of the wealthiest quintile of people have changed how they operate. A revisionist era of "philanthrocapitalism" has emerged. Philanthrocrats, and the structures they support, increasingly link I/ANGOs with private capital to channel their profits into the "developing world", particularly Africa, using corporate social responsibility ideology. Philanthrocapitalism, using business profits from market enterprises to pursue social and environmental goals, has:

"during the past fifteen years, and especially the past five, influenced private foundations to become increasingly directive, controlling, metric focused, and business oriented with respect to their interactions with grantee public charities in an attempt to demonstrate that the work of the foundations is 'strategic' and 'accountable'." (Jenkins 2011, 1)

Some of the central features of philanthrocapitalism include:

- **"quantiphilia"** – a fixation on size, growth, and numbers as indicators of success; data and algorithms have priority over judgment and democracy when decisions must be made (neoscientism), versus the softer mechanisms of socio-politics and human interaction.

- **high engagement by the funder** in the recipient country and NGO implementation;
- **expanding spheres of influence** to advance philanthropic goals for society (norms, behaviours, and policies) to address social problems, including the WASH sector; and,
- **declining acceptance of unsolicited grant proposals** – a "don't call us, we'll find you" attitude that reflects a foundation's belief that they know more about how to solve social problems than NGOs or communities (Jenkins 2011, 10–20).

The main critique of these rich foundations, usually dominated by a single individual (the founder), is that they operate without significant accountability to the public or state apparatus. This aggressive "muscular philanthropy" undermines community sovereignty, negating social change and the building of communities and social ties through empowerment and participation (Jenkins 2011, 17). Drivers of philanthrocapitalism include Bill Gates, Warren Buffett, and Bill Clinton, who shape philanthropic giving through events such as the Clinton Global Initiative and the World Economic Forum in Davos, Switzerland. Philanthrocapitalists (active in) Africa include elite men like Kofi Annan, George Soros, Mo Ibrahim, Bono, Tony Blair, Aliko Dangote, Theophilus Danjuma, Tony Elumelu, Patrice Motsepe, and Tokyo Sexwale (Edwards 2013, 22–29).

While I argue that African ED approaches are a necessary precondition for the right kind of development in Africa, the challenge is how to advance ED and IK without their being manipulated to serve such neo-liberal capitalist interests. Following an overview of ED below, I examine this dilemma in Ghana, based on my practical work in the WASH sector where I daily struggle within an INGO to overcome the contradictions such spaces, dependent on Western aid and philanthrocapitalists, foster.

Endogenous development and culture

The emphasis on cultural approaches to development emerged during the 1982 World Conference on Cultural Policies (MONDIACULT) held in Mexico City, and the subsequent years of the UN World Decade for Cultural Development (1988–97), which acknowledged that cultural imperialism has been the norm of past efforts focused on economic or political development that have failed to reduce poverty in African or other countries in the "South" or "Third World" targeted by the Northern powers. More importantly, Western development projects failed to render indigenous culture totally useless in the minds of Africans (Yefru 2000). A cultural lens offers a critical view in which the West's knowledge claim of what development is can be rightly seen as biased, one that eschews indigenous worldviews and ways of life, and concepts of knowing, doing, and being that constitute their humanity. The global right of indigenous people to live their culture has been reaffirmed through adoption of resolutions at the Third International Decade for the Eradication of Colonialism (2011–20) and the Second International Decade of the World's Indigenous People (2005–14).

Theorising ED by re-conceptualising the role of culture in development and identity construction can provide an avenue to uphold African indigenous people's essential survival imperative and opens the civic space for ED begin to counter the historical hegemonic enterprise of Western development. A definition of culture is important:

"Culture is not simply the song and dance of a people. Nor is it merely the compilation of their holidays or the listing of their heroes and heroines. Culture is a vast structure of behaviors, attitudes, values, habits, beliefs, customs, rituals, language, and ceremonies peculiar to a particular group of people that provides them with a general design for living and patterns for interpreting reality ... A peoples' indigenous culture anchors them to reality and must be the starting point for all learning.
– Nana Kwaku Berko I." (Nobles 1990, 1–12)

The concepts of cosmovision – a social construct that includes the assumed interrelationships between spirituality, nature, and mankind (see Millar et al. 2005) – and ED, as applied to African peoples, centre on an understanding of the symbiotic relationship between culture and worldview. Indeed, the culture of a people encompasses worldview elements that include ontology, gnoseology, epistemology, axiology, cosmology, aesthetics, and teleology. To engage ED for development, and WASH sector interventions in particular, requires understanding the culture and worldview of African peoples that underpin their source of strength, have evolved over millennia, and guide and shape future collective development and aspirations. Development, in this sense, requires knowing and using African indigenous systems of knowledge, science, relations with nature, spirituality, and societal institutions (e.g., family, leadership, education, economic), as the basis for achieving community well-being and dignified, self-actualised development.

To use an ED approach does not mean ignoring the problems in a community because of certain cultural beliefs or practices (especially those that have been adapted to survive conquest), nor ignoring wider systems of structural inequality and power relations at local levels within communities. In particular, I draw attention next to how worldview knowledge is essential for important behavioural changes to improve community health and WASH outcomes. There is a serious need to address the lagging sanitation conditions of nations and communities across the continent of Africa. *Ghana currently has only 14% national coverage of improved sanitation facilities and high incidences of open defecation with sometimes fatal health implications.* I will return to these challenges momentarily. However, I argue that addressing this issue should start with internal community dialogue for solutions. In terms of ED, African culture and worldview (cosmovision), the spirituality of many African people is central to their existence and shapes their concepts of health, disease, and wellness, and how they relate to WASH matters as a result. These relationships must be investigated and brought into focus to address the pitiful state of human and environmental sanitation, not only in Ghana but across Africa. What is to be done by the state and others, including I/ANGOs, is better achieved when an ED approach is used to find solutions to structural inequalities (internally/externally) that in many ways created the health challenges, poverty, and uneven access to WASH services.

Aim of endogenous development

What are the specific aims of ED? Broadly speaking, the aim of ED is to empower local communities to take control of their own development process in a manner that builds on the existing indigenous institutions and resource base of communities, including their natural/material, social, and spiritual resources. This requires using and/or revitalising ancestral and local indigenous knowledge and culture, while selecting those external resources that best fit the local conditions and "surthrival" needs. Figure 1 shows the core components of the ED approach and inquiry, but most importantly shows the interconnectedness of African people's spiritual,

Figure 1.   Central components of endogenous development inquiry.
Source: adapted from www.cikod.org

social, and material world experiences which when seen in isolation gives a distorted view of a community's worldview, beliefs, and behaviours.

ED principles and theories of change

Some of the commonly articulated assumptions and theory of change aspects of ED, considered as emanating from a "strength-based" approach essential to sustainable development, include:[2]

- Endogenous development (process, methodology, vision) is controlled by local community actors.
- People feel confident and energised to move into the future when they can bring with them experiences that have given them a sense of pride about their identity, capacity, and abilities in the past. Furthermore, communities draw power and energy for change from the internal real life stories of past resilience and survival based primarily on their cultural practices and worldviews.
- Meaningful and lasting change always originates from within. In the context of war, conquest, and colonialism (Arab, Western/European) the resulting external changes, in spiritual and religious beliefs for example, are in direct contradiction to this principle of ED. Prior to colonialism, most cultural change practices were voluntary, except in times of internal conflict and subjugation to stronger ethnic groups. The key idea here is that culture is not static, but also should not be imposed.
- Every single person has capacities, abilities, gifts, and ideas. Central to this idea is the notion that some people, in any situation, are getting things right. This supports a community's ability to build confidence and positive self-identity of its members and collectively move forward using the abundance of existing and untapped potential competencies, associations, resources, and assets. In African societies, doing so is normally more from a collective sharing perspective than an individual materialist paradigm.
- Civil society at the rural level is visible in the form of indigenous organisations (*Nnoboa* groups, *asafo* groups, *susu* groups, clan networks, and hometown associations) through which poor rural families organise their lives. The Centre for Indigenous Knowledge and Organizational Development (CIKOD) has had success using ED to improve livelihoods in communities through these organisations. They have also been able to revive, where absent, many of these institutions to drive contemporary development of indigenous communities in Ghana.

Despite historical antecedents of colonial and state displacement of African peoples in order to access and control their land, water, and other resources, the contemporary nature of inadequate and unequal access to clean, safe water and improved sanitation and hygiene services is one of the primary reasons for high levels of disease, poor nutrition, and mortality of African people. Currently 884 million people worldwide are without access to clean water and 2.6 billion are without proper sanitation (UN 2010). This sanitation crisis is the primary cause of diarrhoea, the biggest killer of children, especially those under the age of five.[3] As one wrestles with solutions to the structural issues of global capital inequalities, the political dominance of the West, and continued marginalisation and dispossession of the poor from their lands, among other challenges, what can be done now to save the lives of African children?

Sanitation, culture, and health outcomes

Research on the linkages between culture and a people's beliefs and practices regarding water, sanitation, and hygiene is emerging to help foster improved WASH and health outcomes, and overall human development. There's no debate that people need clean, safe, potable water and improved sanitation and hygiene practices to live, and to eliminate or greatly reduce water-borne diseases and instances of diarrhoea. ED helps to frame the issue of WASH in a more relevant context, and can unravel core cultural beliefs that shape everyday behaviour and practices that hinder or can improve health and life chances. The concept of sanitation adopted here is articulated in research conducted in Nigeria by Akpabio and Subramanian's work (2012). Sanitation:

> " … is related to all aspects of personal hygiene, waste disposal, and environmental cleanliness which could have impact on health. There often exists a linear connection between dirt, water, and disease – covering personal and domestic hygiene, vector control, food cleanliness, and drinking water storage. This is compared to most intervention efforts today that narrowly conceive of sanitation as toilet construction, rather than a package of environmental and household cleanliness, with water assuming a central position." (Akpabio and Subramanian 2012, 1)

Furthermore, sanitation includes the physical environment as well as social, cultural, and temporal factors that operate as fundamental contexts defining knowledge, behaviours, and decisions relating to problems of water supply and sanitation practices (Akpabio and Subramanian 2012, 1–5). Akpabio's research among the Ibibio people in Nigeria challenges conventional thought. He states:

> " … although water mediates the transmission of microorganisms or parasites onto humans, unsafe sanitation practices and lack of environmental hygiene catalyze the spread of infections. Conventional science and knowledge often employ the logic of biological-epidemiological evidence in understanding such environmental and behavioural health perspectives of water contamination and diseases spread (Curtis et al. 2000). This has led to standard categorization of water related infectious diseases transmission pathways. *What seems to be missing in such thinking is the role of cultural factors of beliefs, local knowledge, norms, values and spirituality in influencing the broader contexts of behaviours for which contaminations and diseases spread occur.*" (Akpabio and Subramanian 2012, 5; emphasis added)

This last point validates the earlier claim that disease, illness, and well-being are culturally constructed and explained. Akpabio and Subramanian's research into WASH practices is guided by what he calls the local/traditional ecological knowledge (L/TEK) framework, closely aligned to the ED approach. Both LEK and ED promote a strength-based community development approach designed to empower and build on the existing culture of a community, particularly their indigenous institutions and resources. However, it is instructive to note that the ED approach does not view local indigenous knowledge as "without" scientific merit or rigour, or only as skills-based knowledge, as does the LEK framework (see Millar et al. 2005).

In sum, Akpabio's research among the Iribibio shows that a people's relationship with water in Nigeria is mediated through learnt cultural experience – individual experiences with water make the context of meaning relevant. He generated a typology outlining spiritual and other cultural meanings attached to WASH that was used to craft successful behaviour change messages that shed light on disease transmission pathways, due to open defecation and ingestion of faecal matter by children. WASH and health intervention programmes designed to improve sanitation and hygiene practices cannot solely depend on the Western scientific understanding of disease aetiology.

WASH and ED case study: reflections on WaterAid Ghana's unfolding story

This next discussion outlines the historical development of ED practice in the WASH sector in Ghana led by WaterAid Ghana (WAG), and highlights some preliminary successes and challenges to the ED approach and methodology.

Viewing development as a cultural process, I have used my leadership position within the INGO WaterAid's country programme in Ghana to promote ED as a means to tackle the poor WASH situation (especially sanitation). As with any change process and entry of seemingly new ideas, there was resistance both within the organisation and among existing partners – the intermediaries between WAG and the local communities we serve, who had not engaged ED theory and practice in any meaningful or substantive way. The opening to introduce ED as formal practice in the organisation for WASH programming work came in late 2011, when funding was offered to WAG by a major bilateral donor agency, Australian Government Aid, through the Australian African Community Engagement Scheme (AACES) programme. Interestingly AACES literature indicated support for organisations using strength-based approaches to development, including ED. My initial hesitancy to accept the funding hinged on the fact that I did not have any internal WaterAid donor intelligence on the Australian government's African development agenda. As a political scientist and scholar-activist concerned with global struggles of African and other indigenous peoples, I certainly knew of the imperial relations of Europeans in Australia and the Aboriginal people who had been dispossessed of their land. I was also concerned that the intellectual property rights (Zakiya, Mgbeoji, and Oguamanam 2009) of the IK documented had not been discussed; of less concern, though also important, were the political risks of implementing a programme for which the time frame had already been given by the donors *before* engaging in bottom-up programme design – a frequent methodological challenge of most development interventions. However, ultimately, the opportunity to use ED to address the disappointing results of current WASH sector sanitation practices informed the final decision to move forward, in addition to the internal pressure to accept donor funds in a changing internal funding regime throughout the organisation.

Historically, the WASH sector lacked concerted focus on facilitating improved sanitation and hygiene practices in general, and in Ghana, until a big push came in 2011. Led by WAG and UNICEF, among other major sector actors, global Sanitation and Water for All High Level Meetings started in Washington, DC to address the fact that the WASH Millennium Development Goal was most "off track" of those to be achieved by 2015. Part of this neglect was linked to the preferences of donors and supporters, who like to count people who they have reached by providing water facilities, with less emphasis on sanitation facilities, through direct subsidies. It was recognised at the meetings that this has happened at the expense of sustainability of interventions and lasting improved health and well-being of communities that improved sanitation facilities and hygiene promotion practices bring through behaviour change efforts. Clearly just providing people with water has benefits, and is essential for life. However, more was now recognised as necessary for healthy lives free of disease. In Ghana, six of the top 10 diseases affecting the general population are WASH-related, namely malaria, skin diseases and ulcers, diarrhoeal diseases, acute eye infections, intestinal worms, and anaemia. According to the Ghana Health Service (GHS) health facility data, malaria is the number one cause of morbidity, accounting for about 38% of all outpatient illnesses, 35% of all admissions, and about 34% of all deaths in children under five years However, *some 10,000 infants die by the age of five each year from diarrhoea*. This has been attributed to poor sanitation and hygiene. Furthermore, *50% of malnutrition is associated with repeated diarrhoea or intestinal worm infections as a result of unsafe WASH* (NMCP 2010 Annual Report). Between 3.1 and 3.5 million cases of clinical malaria are reported in public health facilities each year, of which 900,000 cases are in children under five years (Ghana National Malaria Control Strategic Plan 2008–2015).

An exciting opportunity had come to jointly chart a path towards really empowering communities enduring WASH deprivation and injustice. While I did not have any direct interaction with CIKOD,[4] one of the pioneering organisations of ED theory and practice in Ghana, I knew of their existence, and of their success using an ED approach to improve community-based natural resource management, and other areas as documented in the COMPAS Africa reviews of member organisations. Taking these and other issues into consideration, WAG joined forces with CIKOD and tasked them to unravel the important socio-cultural, spiritual, and material beliefs related to WASH among the different ethnic groups in Ghana living in our project communities, in order to improve sanitation and hygiene practice results. The first year of work would require capacity building of staff and partners to build their understanding of the theory and practice of ED.

One of the obvious implications of our shift to ED is that donor requirements for quick "user numbers" of people served by improved sanitation or water facilities built was not going to be necessarily or immediately forthcoming. Using the ED approach, a community's vision for development and WASH technology choices would drive interventions – and they may not choose what we had as a goal – with a focus on WASH issues as a development priority. Another reason for lower sanitation facilities possibly being built towards achieving prized user numbers – is due to a recent shift to nonsubsidy-driven sanitation work.[5] WaterAid UK adopted community-led total sanitation (CLTS), which marked a shift to nonsubsidy-driven provision of sanitation services and facilities. CLTS has a theory of change that communities could be shamed and shocked into moving from open defecation to at least dig and burying faecal waste matter. This rarely happens quickly. Additionally, compared to when both water and sanitation facilities were built with direct subsidy monies, an ED approach shifts decision-making to communities using what existing traditional structures, people, and socio-cultural-attitudinal and behavioural strengths a community has, that may be needed to change and improve poor sanitation and hygiene practices that contribute to disease, ill-health, and possibly death.

The ED approach requires knowledge of the meanings and worldview understandings of WASH issues found within a community. This knowledge is acquired through a slow process of community entry and relationship building to foster sustainable results based on their vision, and how the communities respond to any new meanings and accepted norms of, in this case, best practices guiding access and use of WASH resources and facilities. Frankly, it is highly unlikely that a community would not want to have water immediately available for use when WAG appears in the community. However, the community entry process and resource mapping exercise used by ED methodology in our projects often provokes wider discussions around access, use, and rights to all natural resources needed for development, including water. Traditional leaders are also questioned in public spaces on the external resources available that they have access to for the community's welfare, and so such dialogue begins to unravel structural issues of inequity and inclusion. Additionally, some of the national level workshops WAG and partners used to bring community members and other civil society actors and government officials together, moved discussions towards examining both local and larger global structures of inequity. This opens the space to interrogate the direction of national development that marginalises indigenous culture and life, while failing to ensure the right to basic essential WASH services, especially for those in extreme poverty.

Documenting results

Some of the positive results to date from the AACES ED project implementation have been documented in several forms: significant change stories of improved sanitation and hygiene behaviour change, and two independent reviews – one currently in process as part of the AACES midterm review process, and the other, a midterm review of our country strategic plan 2011–2015 (CSP5) that also assessed ED implementation. An overview of specific outcomes of ED implementation in Ghana includes:

(1) Baseline profiles of communities and their socio-cultural views on WASH. For example, in a workshop in Techiman, Ghana in 2012, participants learnt how important it is to understand the links between spiritual meanings attached to water to get a community to use safer drinking water. In this case, where water from a particular river had spiritual significance, a bore hole was sunk near the sacred river where people had been drinking, as it became clear that the location for drinking the water was the important aspect in the ritual and not that the water came directly from the river itself.

(2) In the case of efforts to improve health outcomes and improve sanitation behaviours that reduce open defecation, our work is uncovering worldview beliefs and meanings and how they manifest around children's health and links to ingested excreta. Taboos related to use of toilet facilities are also being respected and documented to inform facility design and use patterns necessary to create important sanitation and hygiene messages.

(3) At the WaterAid International Programmes Conference in 2013, WAG presented a significant change story of the Okrakwadwo community as an ED success story. In this community, located in the Eastern Region, Ghana, there lives a small population of 1,338 people (women 366; men 329; girls 301; boys 342) made up of diverse ethnic groups (Akan, Ewe, GaDangme, Dagbani, FraFra, Dagari), with two people identified as differently-abled. The "before" situation in Okrakwadwo included poor management of water points, resulting in breakdown of the facilities attributed to the lack of maintenance and transparency and accountability of community resources. Through the ED approach, traditional authorities, indigenous community groups, the Akuapem North District Assembly, and implementing partners (Akuapem Community Development Programme (ACDEP) and CIKOD, engaged in community entry and sensitisation processes that led to increased dialogue and creation of a shared understanding of available resources of the community. Indigenous knowledge of WASH historical practices were documented, and creation of a shared vision of development priorities was produced by the community to chart their own development path before any external support was introduced. The ED project results include: community resources used to extend the reach of the water point by the community, a more inclusive reconstituted Water and Sanitation Management Team, and using an ED tool, the Learning, Sharing, and Assessing (LeSA) platform, sharing their ED experience with other AACES programme communities. One challenge for WAG and partners going forward, however, is to do more in-depth power analyses of the emergent changes.

(4) WAG's midterm review of CSP5 (Millar and Malunga, forthcoming), is currently under review by WAG management. However, the draft report has noted that the introduction of *"the indigenous cultures, practices, and experiences in water and sanitation management of the local people"* into operations is *"highly commendable and should be celebrated"*. Additionally, WAG should share its ED experiences with *"local collaborators and the wider WaterAid family as a best practice"*. Furthermore, the evaluators found that in terms of value-for-money: *"communities where the ED approach is used reflect high and more vibrant returns in terms of levels of participation and identity (identifying with actions and results as their own)."* Some initial resistance by partners has also been reduced as ED *"is beginning to be appreciated by Partners"* (Millar and Malunga forthcoming, 20–32).

WAG has also learnt that for those who wish to promote an African-centred ED approach, it is often the processes employed (methodology and methods) that are far more important than outcomes. Such processes are expected to be decolonised, respectful, and enable people, to heal, to educate, and to lead steps towards self-determination (L.T. Smith 2005, 128).

The future of ED in WaterAid Ghana

WAG is entering the third of a five-year funding period for the AACES programme. The Australian government appears to have changed policy focus, after integrating development with foreign affairs, and future funding will not be available for AACES. How long the ED approach is used in WAG should not be linked to AACES funding or my presence in the organisation, but should have continuity in the face of any leadership changes, if what is being done shows signs of success – as defined by the communities and those within the organisation who have a shared belief in the value of ED to improve WASH outcomes and impact, especially sanitation- and health- related ones. What continues to be important is for our work to support communities to understand their WASH and other practices in a way that respects their cultural significance and speaks to improved well-being and thriving health indicators they have identified, tracked, and deemed better, whether or not WAG is present in their sacred space. Continuous feedback from the community is required. Our partners must hold workshops for community members to share their experiences and challenges from our relationships. The core challenge will be to document the process of engagement, how and if the theory of change embedded in ED is valid, and if ED contributes to the political economy of knowledge production on WASH and cultural development in Africa. Another core challenge will be to demonstrate the impact of ED beyond isolated success stories and to influence policymakers to scale up the approach countrywide.

The anticipated results from using ED in WASH will unfold more concretely in the coming years. Managing the tension between going at the pace of the community and meeting donor time-lines and expectations will be necessary to avoid ED being seen as a quick fix, and to allow organic relations to develop. While this is a concern and political risk to handle, another more important tension arising from using ED in this context is the need to quickly save lives, as every day the statistics on water- and sanitation-related diseases show unnecessary suffering of Africa's children. WASH-poverty kills, and those most marginalised and without the resources to ensure continuous or at least regular access to safe, clean water and sanitation facilities are forced to live in intolerable and inhumane conditions. Clearly, this must end.

As efforts begin to move in positive ways on the ground, gathering the evidence of other necessary changes in the structural systems of inequality that privileges the global rich and private sector actors with access to clean water must be brought to light for concerted collective action because focusing on creating the foundations for good sanitation and hygiene practices at the micro-level, is not enough. Towards this end, WAG will need to re-examine how it works with government, ANGOs and INGOs, and communities through agreements that do not concretely address the structural and power relations (or the lack thereof) of the parties involved, or the real extent to which all actors have the right level of accountability and transparency. Account-ability forums do exists, but need strengthening. WAG should also engage in learning forums on ED, such as the COMPAS network. In the course of 13 years of action research, the COMPAS partners have designed, applied, and tested a variety of methods for enhancing endogenous development, particularly methods for learning from and with local people, for testing and improving indigenous practices, and for networking and training.

Conclusion

This discussion sought to contribute to the discourse of ED practitioners by sharing emerging evidence that the WASH sector can make significant strides in addressing the poor state of sanitation and WASH related incidences of disease in Ghana, and throughout Africa, using ED. It argues that since culture is the foundation of all people's development, the relevance of ED is evident as ED recognises that culture constitutes the totality of a people's existence. Through research findings in Nigeria and ED implementation in Ghana by WAG, the discussion further showed how culture shapes worldviews and informs concepts of health, wellness, and healing found in relation to WASH. As a result, the opportunities to reduce waterborne diseases like cholera, diarrhoea, and others, and to improve child health and nutrition, are limitless in both WASH and wider development aims, when the right approach facilitates organic, community-led solutions in culturally sensitive and appropriate ways. While some indigenous institutions have been weakened from over four centuries of colonial and neo-colonial onslaught, most peoples in Africa, especially in the rural enclaves, still rely on these institutions for sustenance, to create and recreate their indigenous and endogenous worldview, sciences and knowledge, and material and spiritual life-systems.

Understanding the strengths and weaknesses of African ED approaches is demonstrably an important step towards creating a viable body of ED knowledge that embeds sustainable solutions to drive the desired behaviour changes needed for positive sanitation and hygiene outcomes. This paper further asserts that while ED promises to bring about a radical change in the praxis of sustainable development, its purposefulness does not lie in its ability to move Western neo-liberal capitalism forward as a managed and integrated approach among a contesting plurality of ideologies. To make global and local progress in Africa, one must critically examine *who* shapes and defines development theory and practice, and support efforts to shift the flow of knowledge about development from the South to the North. African and INGOs that collaborate to function as comprador institutions remain a stumbling block to Africa's development and must be challenged when they exploit indigenous people and local communities for development aid from their sponsors.

In the final analysis, the use of ED and IK to achieve the seemingly elusive African-centred state of perpetual sustainable development may ultimately require directly challenging, through collective organising, the structure of a society that facilitates WASH injustice and elements complicit in such oppression, including the neo-colonial state. In addition to overcoming the challenges of compradorism and philanthrocapitalism, countervailing forces must prevail, in order to allow for the alternative ways of thought promoted by ED and IK to emerge – the ways in which they allow thinking in decolonised ways regarding orthodox development theory, methodologies, practices, and knowledge. Despite the continued existence of development as an essentially missionary enterprise, the ED approach and processes are expected to be decolonised, respectful, and enable people to heal, to educate, and to lead generations towards self-determination. There is growing knowledge available about the negative results of orthodox development projects, but more research is needed on worldview understandings, and on the power relations embedded in these processes, including protection of indigenous knowledge through appropriate intellectual property rights regimes. These are some of the critical challenges African scholars and practitioners of IK and ED must overcome in solidarity and true partnerships with communities.

As WASH-related ED programming and success emerges in Ghana, the unfolding WAG story needs to be continuously updated, validated by communities, and shared in ED and other progressive networks across the continent, particularly among COMPAS members. However, beyond the WAG story, it will take dedicated and committed Africans to come together to find holistic, comprehensive African solutions to Africa's development opportunities and challenges. While not discussed in detail here, the local and structural causes of WASH injustice faced by the poor and marginalised might not be responsive to these groups *unless* a social movement or shift occurs, led by informed people willing to fight for, and demand from the state, global capitalists and donors (who predominantly still control African economies and resources), their right to a self-determined future. This future must be characterised by equitable and fair access and control over WASH services – not just for people today, but for the Africans of future generations.

When this happens, the influence, power, and control of neo-compradors will decline, wherever they are located, and their ability to manipulate ED and IK for non-African interests will fade.

Notes

1. Non-governmental organisations (NGOs); these are considered private, voluntary, not-for-profit organisations not associated with the state or multilateral or bilateral institutions. Three types are discussed in this article: international NGOs (INGOs) with headquarters in Europe, America, or another continent, and usually with locally-based arms in Africa or other developing continents. I refer to INGOs as northern NGOs (NNGOs), denoting the geographic origin and power relations of the organisation (internally and externally). The next type of NGOs are those that are an extension of the parent INGOs, the local chapters operating in Africa (IANGOs). The last NGO type discussed are locally-originating NGOs that do not have global locations but may have several national offices. These NGOs in this article are called Southern NGOs or African NGOs (SNGO/ANGO). Most of this piece critically examines the unequal power relations and flows of 'solutions' to development between both the INGO headquarters and local chapters, and the local chapter and ANGOs down to communities.
2. Part of this section is adapted and extracted from a power point presentation of the AACES, Australia-Africa, a Community Engagement Scheme programme in Nairobi, December 2010. The AACES programme provided the entry point for me to introduce ED into WaterAid Ghana programming, which, to my knowledge, is the only WaterAid country programme that is integrating ED and IK theory and practice in WASH.
3. Other fatal and debilitating WASH-related conditions include under-nutrition, trachoma, worm infections, and cholera.
4. The Center for Indigenous Knowledge and Organizational Development (CIKOD, www.cikod.org); while I do not necessarily see CIKOD as a neo-comprador NGO/organisation, this is yet to be fully explored. Its mission and role in pioneering ED in Africa and fairly progressive agendas in Ghana sets it apart from most other African NGOs. It is a part of the COMPAS network of ED practitioners. Our other local NGO partners include Pronet South and ACDEP, among others.
5. The Government of Ghana, MLGRD/EHSD mandated community-led total sanitation (CLTS) as the primary approach to sanitation scale up in Ghana in 2011 under Kamal Kar; it has achieved mixed results (http://www.cltsfoundation.org/).

References

Akpabio, E. M. 2012. "Water Meanings, Sanitation Practices and Hygiene Behaviours in the Cultural Mirror: a Perspective from Nigeria." *Journal of Water, Sanitation and Hygiene for Development* 2 (3): 168–181.

Akpabio, E., and S. V. Subramanian. 2012. "Traditional Ecological Knowledge: An Emerging framework for Understanding Water and Sanitation Practices in Nigeria." ZEF Working Paper Series 94. Bonn: Zentrü für Entwicklungsforschung, University of Bonn.

Aubrey, L. 1997. "Moving Beyond Collective Learning from the North and Bringing Humanity Back to Itself: PanAfricanism, Women, and Co-Development," Paper presented at Panafest Colloquium in Cape Coast, Ghana, September 1–5.

Edwards, R. "The Philanthrocrats: Doing Good at a Price." *The Africa Report*, 1 February 2013. Accessed May 28, 2014. http://www.theafricareport.com/North-Africa/the-philanthrocrats-doing-good-at-a-price.html

Fernando, J. L. 2003. "NGOs and Production of Indigenous Knowledge Under the Condition of Postmodernity." *The Annals of the American Academy of Political and Social Science* 590 (Nov.): 54–72.

Fowler, A. 2000. "NGO Futures: Beyond Aid. NGDO Values and the Fourth Position." *Third World Quarterly* 21 (4): 589–603.

Hearn, J. 2001. "The 'Uses and Abuses' of Civil Society in Africa." *Review of African Political Economy* 28 (87): 43–53.

Hearn, J. 2007. "African NGOs: The New Compradors?." *Development and Change* 38 (6): 1095–1110.

Jenkins, G. 2011. "Who's Afraid of Philanthrocapitalism?." *Case Western Reserve Law Review* 61 (3): 1–69.

Millar, D., and C. Malunga, forthcoming. *WaterAid Ghana (WAG), Light-Touch Mid-Term Review of CSP5 (2011–15)*, final report.

Millar, D., S. B. Kendie, A. A. Apusiga, and B. Haverkort, eds. 2005. "African Knowledges and Sciences: Understanding and Supporting the Ways of Knowing in Sub-Saharan Africa." Book of Papers and Proceedings of an International Conference on African Knowledges and Sciences, Bolgatanga U/R Region Ghana, 23–29 October; downloadable from COMPAS. Accessed May 28, 2014. http://www.compasnet.org/blog/?page_id=599

Mohan, G. 2002. "The Disappointments of Civil Society: The Politics of NGO Intervention in Northern Ghana." *Political Geography* 21 (1): 125–154.

Nobles, W. 1990. "The Infusion of African and African-American Content: A Question of Content and Intent." National Urban Alliance. Accessed May 28, 2014. http://www.nuatc.org/articles/pdf/Nobles_article.pdf

Rodney, W. 1971. *How Europe Underdeveloped Africa*. Washington, DC: Howard University.

Sachs, W. 2010. *The Development Dictionary: A Guide to Knowledge as Power*. New York: Zed Books.

Smith, L. T. 2005. *Decolonizing Methodologies: Research and Indigenous Peoples*. New York: Zed Books.

UN. 2010. "MDG Gap Task Force Report 2010: The Global Partnership for Development at a Critical Juncture." Accessed May 28, 2014. http://www.un.org/millenniumgoals/pdf/GAP_FACTS_2010_EN.pdf

Yefru, W. 2000. "The African Challenge to Philosophical Paradigm: The Need for a Paradigm Shift in the Social, Economic, and Political Development of Africa." *Journal of Black Studies* 30 (3): 351–382.

Zakiya, A., I. Mgbeoji, and C. Oguamanam, eds. 2009. "Proceedings of the UI IK Study Group." Intensive Workshop on African Indigenous Knowledge and Intellectual Property Rights: Implications for Nigeria's Development. University of Ibadan, Nigeria, 20–23 April 2009. Accessed May 28, 2014. http://issuu.com/profibadan/docs/final_proceedingsaug09

Endogenous development in Somalia: bridging the gap between traditional and Western implementation methodologies

Ariel Delaney

This practical note examines the implementation approach of African Development Solutions (Adeso) in Somalia, a country which is recovering from over two decades of conflict. It discusses how their endogenously derived targeting method, known as ICBT, is implemented and the way it challenges social norms for positive outcomes. Cash-based response is analysed as a recovery method as well as a way to engage community participation, particularly with marginalised groups. Implementation challenges are highlighted to explore the relationship between traditional and globalised (Western) values.

Cette note pratique examine l'approche de la mise en œuvre d'African Development Solutions (Adeso) en Somalie, un pays qui se relève actuellement de plus de vingt ans de conflit. Elle traite de la manière dont sa méthode de ciblage d'origine endogène, connue sous le sigle ICBT, est mise en œuvre et de la manière dont elle met en question les normes sociales pour des résultats positifs. La réaction basée sur l'argent est analysée comme méthode de relèvement, ainsi que comme manière de stimuler la participation communautaire, en particulier parmi les groupes marginalisés. Les défis pour la mise en œuvre sont soulignés pour examiner la relation entre les valeurs traditionnelles et mondialisées (occidentales).

Esta nota práctica examina el enfoque dado a la implementación de Soluciones de Desarrollo Africanas (Adeso) en Somalia, país que se recupera de más de dos décadas de conflicto. Considerando este contexto, analiza cómo ha implementado su método de selección, generado endógenamente y conocido como ICBT, y cómo el mismo cuestiona las normas sociales existentes en torno a los resultados positivos. En este sentido, analiza la respuesta obtenida a partir del otorgamiento de dinero en efectivo como método de recuperación y de incentivo de la participación comunitaria, particularmente en el caso de grupos marginados. Esta nota destaca los retos implicados en la implementación, como manera de explorar la relación entre los valores, tanto tradicionales como mundiales (occidentales).

Introduction

Over the past decade or so the global south has seen the emergence and growth of local non-governmental organisations. Many of these organisations use different approaches to implement their development initiatives. African Development Solutions (Adeso) is an endogenous organisation that believed some Western participatory methodologies were not effective for the local context in which they primarily operated, Somalia. Adeso's founding belief is that development must come from within, not from outside African communities and that Africans must determine Africa's future. Fatima Jibrell, a Somali-American environmental activist, founded Adeso in the midst of the chaos that was unfolding in Somalia in the early 1990s. Adeso operates in a way that aspires to preserve African culture and values because it believes that *"while international aid has provided much-needed support, it often falls short of enabling lasting change at grassroots level"* (Adeso 2012a). Adeso wants to change this and feels that their strong bonds with African communities and Adeso's belief in the capacities of the people enable them to do so.

Recovering from over 20 years of conflict and without a central government, Somalia is a fragile state. Its population is one of the most vulnerable in the world. Unpredictable weather patterns and severe environmental degradation complicate efforts for development. Livelihoods are under stress, especially in rural areas that lack basic necessities for daily living. The role of the international humanitarian sector is vital in the day-to-day lives of the population.

Adeso works with vulnerable communities on emergency and development issues. The organisational focus is to save lives, rebuild and protect livelihood assets while working across programme areas to reinvigorate local economies, provide humanitarian aid, develop skills for life and work, and influence policy. Programming is driven by the belief that economic, social, and environmental security is the foundation of a healthy community. Adeso has been working in Somalia for 20 years, and has a dedicated and experienced staff of primarily Somali nationals implementing its projects. It also has field operations in Kenya and South Sudan.

Adeso's approach

As an implementing agency and a major international non-governmental organisation (INGO) working in the Horn of Africa, Adeso utilises the participatory methodological approach at the community level referred to as inclusive community based targeting (ICBT) for project implementation. The ICBT methodology has been used intensively and successfully by Adeso in the context of Somalia since its inception in 2003, and is aimed at targeting the most vulnerable populations through a community participation process. The ICBT process includes community mobilisation, community based selection of beneficiaries, public verification and validation of beneficiary list and establishing village relief committees (VRCs) to implement projects with full community participation. Adeso's interventions and ongoing monitoring, utilising the ICBT approach, show evidence of improving the lives of the Somali population they serve. Prior to the implementation of the ICBT methodology, extensive assessments are done to determine the overall situation of a given region. For example, these analyses can include: situational, environmental, gender, food security, livelihoods, and market factors. Adeso develops the assessments based on current information from various networks. Transparency, empowerment, gender awareness, and community participation are Adeso's ICBT guiding principles for effective targeting and project implementation.

Adeso believes successful projects result from quality stakeholder engagement, research, insight, careful planning, implementation, and constant adjustments based on feedback provided by a monitoring and evaluation system for each and every stage of the project life cycle. All projects are approached from a project life cycle perspective, following a series of interdependent stages. Typically, these stages include assessment, planning, implementing, and evaluation/ concluding. ICBT begins at the assessment stage and continues until after the projects conclusion via post-intervention M&E (see Figure 1).

Process execution solely depends on the project goals. For example, a primary component of a project intervention is cash-based response (CBR) programming, which Adeso pioneered in Somalia. If a cash-for-work (CfW) project is designated, community members are asked to identify the specific project, which depends on the community's own needs and desires.

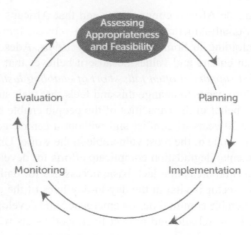

Figure 1.   Adeso project cycle (including M&E). Note. Adeso. 2012a, 2012b. A Practical Guide to Cash-Based Responses.

Common CfW projects are: *berket* construction for water access,[1] road construction for market access, etc. During the ICBT process, the community is assembled for a public forum, which is typical in Somali culture. Adeso's staff is predominately Somali, including many locals, and comprised of men and women of various clans that oversee the community during this process as they know the culture and language. Once the village relief committee (VRC) is agreed upon and reflects Adeso's requirements of equal gender and clan representation, another community committee is constituted to serve as an overseer of the VRC. The VRC mimics the traditional Somali system of the elders participating in communal problem solving but does so by way of ensuring equal community representation. Adeso's approach of an additional committee to oversee the VRC as a monitoring/accountability mechanism shows that they do not entirely accept all existing local social and cultural norms but use them as a starting point for implementation. The VRC ensures that the beneficiaries listed are in fact the most vulnerable in the community. While projects may work in specific sectors, an entire community may not be directly targeted for involvement in the project specifically. That is not to say that they are not impacted indirectly. For instance, a *berket* will benefit everyone but it may be that only a small group of people are involved in its construction. All ICBT steps, especially targeting, must be undertaken with the community as a whole under the guidance and support of the VRCs, while Adeso staff facilitates the VRCs' process. The overall process of targeting, regardless of project type, is to ensure that programmes benefit the intended population or groups within the population (Adeso 2012b, 53).

Adeso believes that there is a *"common and recurrent challenge in implementing humanitarian programs by international aid agencies [which creates a] gap between theory and practice, especially at the field level"*. Targeting is a complex and time-consuming process that must be conducted with transparency and community involvement.[2] (2012b, 54)

Most agencies aspire to use community-based approaches, but this is difficult in practice. The on-the-ground situation in Somalia is *"politically, culturally and logistically challenging to materialise at the field level, especially in emergency situations and for agencies that do not have a strong presence in the field"* (Adeso 2012b, 54). As a whole, communities are not sufficiently consulted or informed about what resources are accessible to them from other agencies or organisations, or even about essential project details, such as the intervention duration or what/who will benefit. Without this information, project impact can be compromised.

Turning vision into practice

Cash-based response

In Somalia, the traditional local authorities are the elders, and meeting with them is crucial. They are the traditional peacemakers and guardians of the community, and as leaders who know the needs of their people, may convince international humanitarian staff that they will transfer the resources fairly to their people. In Somali society value is placed on benefiting the community as a whole rather than the individual. An important example can be noted from Adeso's cash relief implementation. When the most vulnerable beneficiaries are given their cash relief payments they often contribute to community *Sadaqa* used to benefit their community.[3] However, exceptions or gaps do exist as they do in all societies and there are some instances where individuals exploit their communities for their own benefit. Unfortunately, in some cases, the elders may be under pressure from their families or other clan members to request prioritisation. Adeso also acknowledges the possible inherent bias that the elders have with regards to community interests, as often they represent one gender (male) and possibly a clan majority. Consequently, this is not to say that they should not be consulted, but it is not *"sufficient for humanitarian staff to allocate the task of targeting the most vulnerable or distribution of resources to one group of the community, such as the elders"* (Adeso 2012b, 54). Adeso notes that this has often happened in times past: cases were identified where humanitarian staff received lists of target beneficiaries where many truly vulnerable beneficiaries were identified, but also many family members of elders who did not fit the criteria were included. It has also been reported that lists have included names of beneficiaries that do not exist, so-called "ghost beneficiaries". In other cases, humanitarian staff have been reported to collaborate with elders in order to get a benefit in return. Elders are now most often Adeso's greatest advocates at the community level as they have seen the improvement from the interventions over time. This has led to their acceptance of the VRCs and the ICBT process as a whole. ICBT, via the VRCs, targets those who are the most food or livelihood insecure, including marginalised groups, women, men, elders, youth, girls, and boys. ICBT helps in building trust between the implementing agency and the community, ensuring accountability to beneficiaries and minimising the risks. As an indirect result this process builds intra-community trust. These are just a few of the possible issues that can arise during implementation that Adeso has addressed contextually. The ICBT methodology is not a 100% guarantee of avoiding the risks, but if its process is adhered to, the intervention constantly monitored, and the minimum training of all project staff takes place (not just the managers), such measures can significantly mitigate the risks.

During times of shock, coping mechanisms are utilised. Traditionally in Somalia, the provision of credit is an essential coping mechanism to address the consequences of drought, lack of income sources, poor livestock conditions, or lack of alternative livelihood opportunities. Pastoralists and other households commonly go into debt to meet their basic food and non-food needs. Repayment is done once livestock is sold, or when crops are harvested. High levels of debt are frequent, in part due to low purchasing power caused by loss of income/assets or from increased expenditures and from inflation in the rising costs of food and non-food items. In turn, this can lead to a collapse of the credit system and force shopkeepers out of business. When poor households are unable to access credit for purchasing food they first reduce the amount of non-staple food items that are consumed, namely meal reduction and smaller portions. In turn, all household spending, including healthcare and education costs, are reduced as well. Another mechanism is sending children to stay with relatives in other villages or cities. In dire situations, such as conflict and environmental crisis (i.e., drought and floods), pastoralists (who are excessively vulnerable) become internally displaced persons (IDPs) when they must move to areas where humanitarian aid is present. The final and most devastating effect for livelihoods is that pastoralists drop out of pastoralism and settle in towns or IDP camps, becoming dependent on humanitarian aid and diaspora remittances. When pastorialist households lose all their assets, including pack camels, they become unable to migrate during periods of crisis.

Adeso utilises cash-based response (CBR) programming as a mechanism of project implementation. This is a large component present in all projects, particularly targeting pastoralists and other livelihood groups. Prior to the initiation of CBR, the implementation of food aid programming was commonplace in the humanitarian sector. Up until 2011, most international humanitarian aid provided was in kind, manifesting as food, seeds, tools, medicines, shelter materials, and household goods (Harvey and Bailey 2011, 11). Food aid has over time been criticised because, in addition to feeding the hungry, issues of amplified food aid have been linked to aid dependency, a decrease in local food production, and changes in consumption patterns. CBR in implementation utilises various combinations of cash relief (CR), and cash-for-work (CfW). Adeso initially used CBR in 2003, in response to a severe drought (Adeso 2012b, 9). Its use continues today in order to address chronic food insecurity and the continuing humanitarian crisis in Somalia. The immediate result is often seen to increase the purchasing power of project participants to enable them to meet their self-identified needs.

Adeso project staff undergo training in CBR. Adeso signs a Memorandum of Understanding between them and the VRCs to ensure that the targeting process is fair. When implementing CBR Adeso never distributes money directly to beneficiaries. A contract is usually drafted between Adeso and a Hawala (a local/national money transfer company) for the financial services rendered. In the absence of banks in Somalia, Hawala locations are crucial for the population to receive remittances from abroad and are also used in CBR. Hawalas distribute to beneficiaries that have Adeso-designed identification cards (see Figure 2), produced to limit fraud. Beneficiaries must present their ID card in order to verify their identity to receive their cash payments.

In Somalia, the implementation of micro-projects facilitates communities to rebuild infrastructure themselves and can contribute to disaster preparedness for future emergencies. Rock dams are an example of this, and have proved to contribute to disaster risk reduction significantly. CfW projects, tree plantings, and the construction of *gabions* to address erosion, result in the creation of sustainable community assets and the fostering of pasture rehabilitation for pastoralists.[4] Adeso's role has proved critical in the recovery phase in revitalising local economies, particularly through the re-opening of credit lines which had previously collapsed.

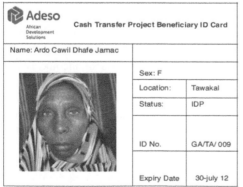

Figure 2. Example Adeso Beneficiary Identification Card (front and back). Source: Adeso. (2012a, 2012b) A Practical Guide to Cash-Based Responses.

Engaging women and marginalised groups

Traditionally, Somalis are very sociable people. ICBT uses cultural and religious norms to enhance participation and to improve the accountability and sustainability of programmes. Adeso builds upon the pre-existing social structures of the elders and community leaders. Key players of each village aid in the formation of the VRCs. However, Adeso encourages a minimum representation of 40% women in the VRCs and encourages minorities to be part of the VRCs. At the same time potential beneficiaries are given information about the project, its objectives, and procedures. This serves as preparation for public meetings (*kulan*), which are grounded in cultural norms. At the *kulan*, the names of the beneficiaries are read out, and either confirmed or rejected by the community at large. The culture is traditionally accustomed to publicly discussing issues and coming to agreement, even though this can take days. Thus Adeso's ICBT approach uses a cultural practice already in place. VRCs are required to take this list of beneficiaries, publicly confirmed in the *kulan* and go house-to-house to verify the families and determine if they are the neediest or selected as a result of favouritism. VRCs, coming from the community, know the context better than Adeso staff members. Although time consuming, beneficiary verification is particularly empowering for communities and creates a sense of ownership. Important teambuilding skills are utilised, particularly negotiation. There are many examples of VRCs becoming very vocal when they believe that needier people have been left out and that relatively wealthier people have been included on the beneficiary list. By promoting the inclusion of women and minorities, Adeso pushes the margins of existing cultural systems. Verification of the beneficiaries also includes another long task of data collection and biographic information from selected beneficiaries. This is done publicly with the VRC.[5] Concerns can be raised by local authorities or community members, and at such times the registration process is halted for more dialogue with and between the communities. In this context full community participation is key.

In 2006, Adeso initiated a women's literacy programme in Sanaag. During the project's mobilisation some leaders came out against the initiative saying that the project was against Islam. Senior project staff member Khadra Yusuf used Koranic verses on rights and gender equality to orient the community to what the Holy Koran actually says. Afterwards, a senior religious leader (*Sheik*), added and explained that Islam encourages the education of women. This is a prime example of the contextual fusion of religion and culture. Since many believe, in general, that the Koran is the best teaching approach, and by promoting education and showing respect towards women, women are more apt to speak up during public meetings than in the past. Many elders now recognise the important role women play at the community level.

Initially with CfW projects, many communities believed that women could not work as well as men for CfW implementation, and therefore were left out intentionally by the VRCs for such projects. Staff adjustments were designed to create labour opportunities for the inclusion of women, and many women participated. After this was challenged, many women advocated being included in CfW activities and Adeso has lead the way in designing projects to incorporate all community members, including women. As a result women have become more informed about issues of marginalisation based on gender, and are slowly fighting for inclusion and recognition in the community development agenda (Chirchir, Odhiambo, and Farah 2012, 29).

Adeso also has a complaints system in place that is available for the community members to call or visit the office at any time if they are unhappy or notice suspicious behaviour. Culturally, Somali women do not make complaints in public, making this of great value for them to communicate with Adeso. This mechanism is monitored even after project completion, and makes for more transparency. Market data is also monitored to track the effect of the cash injections on the local economy. Somali staff and women are key to changing norms. In addition, regarding the effect of the ICBT approach on gender norms, Adeso itself has contributed to the shifting of social norms regionally. Approximately 98% of the field staff is Somali and Muslim. The culture is sensitive in terms of religion, and as the majority of staff are from the culture, they often have a better understanding of the on-the-ground dynamics. Senior staff include both men and women, and gender mainstreaming is part of the ethos of Adeso within the organisation and all operations. This is significant because before Adeso existed in Somalia, having a female-headed organisation, not to mention hiring senor ranking female staff as well, was simply unheard of; furthermore, a Somali female. This demonstrates that culture, with its norms and practices, is not stagnant and evolves with time. By hiring female staff, this indirectly demonstrates to Adeso's female beneficiaries a positive female role model.

Conclusion

One of Adeso's major achievements has been ensuring sustainable food security. The provision of seeds and tools to agro-pastoralists has decreased negative coping strategies and increased the amount of meals consumed among beneficiaries and their families. The data from Adeso's reports suggest that projects revitalise local economies though cash transfers (from post-market monitoring after cash transfer distributions) (Chirchir, Odhiambo, and Farah 2012, 23). Cash programmes change beneficiaries' lives. Many INGOs are now transitioning to cash and Adeso is using lessons learnt from distributions to lead a working group on cash, which is comprised of area NGOs and INGOs, based in Nairobi, where monthly meetings are held to share experiences and ideas. Most beneficiaries, regardless of gender, use cash for repaying debts, basic needs, and education/healthcare costs. INGO interventions must be able to accommodate the conditions on the ground and merge economic development approaches for reducing poverty with disaster risk reduction while incorporating a conflict sensitive approach, which supports and grows traditional conflict prevention mechanisms within Somali culture.

Adeso, like all INGOs operating in the region, has its own struggles in making change in Somalia but works towards addressing the needs of the population, particularly women.

Adeso's approach as "endogenous"

Adeso's unique implementing approach, which was derived from traditional Somali community leadership structure, solidifies it as endogenous. VRCs link and fuse traditional governance practices with the development process. VRC members demonstrate that community members are active participants in their own development. While implementation challenges highlight the relationship between traditional and globalised (Western) values, Adeso's methods have potential for scaling up and adaptation in other contexts.

Notes

1. A *berket* is a traditional Somali water catchment system.

2. In the case of a situation where the entire community is in dire need of assistance, such as a after a rapid onset of an emergency like a flood, all households may have lost their possessions and are in need of assistance. In these circumstances, Adeso uses universal targeting.
3. *Sadaqa* is Islamic tithing.
4. A *gabion* is a cylinder, bottomless basket or even metal framework that appears in a variety of sizes that are filled with rocks and used in civil engineering projects to control soil erosion, especially gulley erosion.
5. Information includes: name, age, spouse, family size, children under the age of five, livelihood/income source, livestock holdings, reasons for settling, mobile phone ownership, and mobile money transfer registration.

References

Adeso. 2012a. "About us." Accessed July 22, 2012. http://adesoafrica.org/about-us/

Adeso. 2012b. *A Practical Guide to Cash-Based Responses*. Nairobi: Adeso.

Chirchir, P. K., A. Odhiambo, and J. Farah. 2012. *Somali Emergency Response Project II Final Evaluation Report*. Nairobi: Adeso.

Harvey, P., and S. Dailey. 2011. "Cash Transfer Programming in Emergencies, Overseas Development Institute #11. Good Practice Review, Humanitarian Policy Group." Humanitarian Practice Network.

Water tariff conflict resolution through indigenous participation in tri-water sector partnerships: Dalun cluster communities in northern Ghana

Sylvester Zackaria Galaa and Francis Issahaku Malongza Bukari

The paper examines how the perceived ownership of a water resource negatively influenced local people's attitudes towards payment of potable water tariffs and exacerbated conflicts between water users and service providers in the Dalun community of the northern region of Ghana. The paper presents a case study of how community participation and endogenous approaches to conflict resolution contributed to payment of water tariffs. The findings show that the establishment of a tri-water sector partnership (TWSP), consisting of the Ghana Water Company Ltd (GWCL), private sector development practitioners, and community water boards, led to positive mediation of water tariff conflicts using the institution of chieftaincy. Alternative, endogenous conflict resolution methods combined with modern methods in a tripartite partnership showed promise as an approach to managing conflicts in water projects and in broader poverty reduction efforts.

Cet article traite de la façon dont la propriété perçue d'une ressource en eau a exercé une influence négative sur les attitudes de la population locale à l'égard du paiement du prix de l'approvisionnement en eau potable et exacerbé les conflits entre les utilisateurs de l'eau et les prestataires du service au sein de la communauté Dalun, dans la région du nord du Ghana. Cet article présente une étude de cas de la manière dont la participation communautaire et des approches endogènes de la résolution des conflits ont contribué au paiement de l'eau. Les conclusions montrent que l'établissement d'un partenariat tripartite dans le secteur de l'eau (*tri-water sector partnership* - TWSP), composé de la Ghana Water Company Ltd (GWCL), de praticiens du développement du secteur privé et des directions de l'eau au niveau des communautés, a abouti à une médiation positive des prix de l'eau en utilisant l'institution de la chefferie. Des méthodes de résolution des conflits alternatives et endogènes, conjuguées à des méthodes modernes dans le cadre d'un partenariat tripartite, se sont révélées prometteuses comme approche de gestion des conflits dans les projets d'approvisionnement en eau et dans le cadre d'efforts plus larges de réduction de la pauvreté.

El presente artículo examina la forma en que la percepción de la propiedad de un recurso de agua por parte de una comunidad dalun en el norte de Ghana incidió negativamente en las actitudes de las personas a la hora de pagar las tarifas de agua potable, agudizando los conflictos entre los usuarios y las empresas proveedoras del servicio. Se presenta un estudio de caso que analiza la manera en que la participación comunitaria y los enfoques endógenos aplicados a la resolución de conflictos ayudaron a facilitar el pago del agua. Los hallazgos en este sentido dan cuenta de que la creación de una alianza tripartita en el sector agua (TWSP) —integrada por la Empresa de Agua de Ghana (GWCL), los operadores de desarrollo del sector privado y las juntas de agua locales— facilitaron una mediación positiva de los conflictos surgidos a raíz de las tarifas de agua, utilizando para ello la estructura de los cacicazgos. La combinación de métodos endógenos de resolución de conflictos con métodos modernos ejemplificados en la alianza tripartita, resultaron prometedores como enfoque a aplicar al manejo de conflictos en los proyectos de agua y en las acciones más amplias orientadas a la reducción de la pobreza.

Introduction: understanding the Dalun cluster communities water conflict[1]

Coser (1956, 8) defines conflict as:

"a struggle over values and claims to scarce status, power and resources, a struggle in which the aims of opponents are to neutralize, injure or eliminate rivals".

Fink (1968, 456) also defined conflict as:

"any situation or process in which two or more social entities are linked by at least one form of antagonistic psycho-logical relation, or at least one form of antagonistic interactions."

The element of antagonism spells the negative character of conflict, but in the case of antagonistic *psychological* relation, Fink (1968) was referring to the fact that incompatibility of interest at the emergence stage of a conflict is normal, and when this remains merely psychological rather than expressed in violent interactions, it is likely that some level of collaboration or compromise could result in a manner that leads to solutions of merit to both parties initially in conflict. Although the notion of conflict often carries negative connotations, conflict in community development issues has been found to stimulate vital changes to the benefit of all stakeholders in development interventions (Coser 1956; Fink 1968; Botchwey 2006).

In this examination of the water tariff conflict in the Dalun-Tamale Tri-Water Sector Partnership (TWSP) Corridor, Coser's definition appears to have an appreciable level of contextual relevance as the indigenes make a claim for the ownership of a water resource, while the Ghana Water Company Ltd (GWCL) makes a claim for the value addition that improved the usefulness of the resource. The situation remained a non-violent conflict, but some form of psychological antagonistic relationship existed between the indigenes and the GWCL as expressed by Fink (1968). This case study is an examination of how collaboration and compromise, through indigenous and modern conflict resolution approaches, overcame the situation after the TWSP intervention.

There is a significant divergence between the common pool resource orientation of the local people and the need of the GWCL to recover costs. Local beliefs are greatly influenced by traditions, such as that *water is a free gift of nature which cannot be sold*. The GWCL and the public-private partnership that evolved has the mandate to recover the cost of value added by the public investment in some basic utility services such as potable water provision. This has been a constant cause of conflict between communities and service providers over issues of tariff determination and payment. This paper examines how ownership of the source of water by communities along the Dalun-Tamale TWSP Corridor, which serves as the water source for the GWCL, generates conflict between those communities and GWCL officials.

Following the water systems rehabilitation and extension project in the Tamale Metropolis and neighbouring communities in 2007 (the Tamale Water Optimization Project [TWOP], an aid project managed by Vitens Rand Water Services BV of The Netherlands), the high cost of the project, about GBP£45 m, led the donor to initiate community participation in the management of the water system along the corridor. The aim was to ensure a reasonable level of payment of water tariffs in poor communities, while sustaining the functioning of the water system through a community ownership approach.

With this pro-poor service initiative, it became necessary to determine the exact causes of conflict in public water delivery services along the Dalun-Tamale corridor, the approaches that were previously used to resolve the conflicts, and the outcome of those approaches. A further stage focused on the assessment of the effectiveness of the TWSP model over the previous water tariff conflict resolution process. The fact that the TWSP comprised local and formal sector institutions which sought to achieve a common goal meant that there was the potential for combining indigenous and modern conflict resolution processes towards the achievement of fairness in water tariff rate setting and collection.

The focus was on how traditionalism impinges on indigenous people's willingness to pay for water, and how the juxtaposed relationship between indigenous and modern conflict resolution approaches helped to fashion an amicable solution to water tariff conflicts in the Dalun community in the northern Ghana. The paper concentrates on instances where local community people are expected to pay for services from public sector projects that depend on natural resources from the local environment; the *interest conflict* between local people and service providers stems from the former's claim of resource ownership (SNV 2008) and the latter's requirement for profitability or cost recovery, after value addition.

The Tri-Water Sector Partnership (TWSP) model

The TWSP model was initiated in 2007, as a result of donor conditionality. The formation of the TWSP was facilitated by the Netherlands Development Organization (SNV) as a contract awarded by Biwater Ltd, the contractor of the TWOP under the directive of the donor. In the partnership, the GWCL, which represented the public sector, owned the assets as well as the provision of services through the use of management expertise, through the use of a public ownership and private operation-type of management contract with Aqua Viten Rand Ltd (AVRL) of The Netherlands (Bukari 2011). The GWCL was also to cooperate with the private sector development actors and community-based civil society organisations.

On the part of the private sector, SNV framed the composition and facilitated the formulation of the partnerships' objectives and activities, as well as supporting its proper functioning by drawing in other local development actors, such as Pragmatic Outcomes Incorporated (POI), which was responsible for the formation and capacity building of the Dalun water boards that replaced the community committees (Bukari 2011). The water boards or civil society represented the community in the TWSP. Their role was to organise and enable community members to reflect on and adjust their attitudes towards water tariff payment and collective maintenance of the public water system.

The Business Partnership for Development (BPD) (2008) considers tri-water sector partnerships as strategic examples of partnerships involving business, civil society, and the government, working together for the development of communities around the world. Successful results from TWSPs yield situations where communities benefit, governments serve more effectively, and private enterprise profits are assured (BPD 2008), resulting in win-win conditions (Botchwey 2006).

Though mainstream literature often proclaims the strength of TWSPs in finding solutions to many development problems, very little information is available about practical results. This paper also attempts to bridge the gap between theory and practice in relation to the newly emerging TWSP model, by focusing on the examination of the causalities of conflict between the indigenes of the Dalun area and the GWCL over issues of water tariff, and how the situation was resolved by the combination of indigenous and modern conflict resolution approaches through a TWSP model leading to increase in tariff collection and rebuilding of trust among the parties.

Methodology

The paper used a before-and-after study design (Kumar 1999). This involved the assessment of the water tariff conflict and conflict resolution situations before and after the TSP intervention in the Dalun-Tamale TWSP corridor. Based on a number of factors such as ownership of the main source of water to the GWCL as an indigenous environmental resource, evidence of the activities of community water boards established by the TWSP, use of available water tariff records for 2006 and 2008 (periods before and after the TSP intervention), and reported cases of water tariff conflict, pilot zones in the Dalun areas including Dalun Station, Dalun Nayili Fon, and Dalun Kanbong Naa Fon were purposively selected as sample sites. This enabled the researchers to make generalisations about the TWSP model in terms of water tariff conflict and how indigenous and modern conflict resolution approaches collectively addressed the situation.

As a control, some communities were studied, like Gbolo and Sorugu, which are also located along the corridor, but not part of the communities that consider the Nawuni Dam (which is the main source of water to the GWCL) in Dalun as their environmental resource, nor included in the TWSP pilot communities along the corridor.

Both quantitative and qualitative research approaches were used. The quantitative approach utilised tariff and water payment data (Kane 1995; Twumasi 2001) in the presentation, interpretation, and analysis of the data on the debit and credit balances of water tariff records before and after the TWSP period, as well as establishing the significance of the TWSP interventions on water tariff payment performance of the communities statistically. This approach made the comparison and contrasting of findings easier.

The use of qualitative research methods, particularly focus group discussions, allowed researchers to get a perspective on how participants perceived the experience of integrated traditional and modern conflict resolution approaches, to determine how meanings were formed through culture, and to discover patterns largely on the basis of narrations, descriptions, and explanations rather than relying only on quantitative variables (Corbin and Strauss 2008).

Results and discussion

The Nawuni Dam in Dalun as the source of water to the GWCL

Dalun is a town in the Kumbungu District of the northern region of Ghana. It is situated 30 km to the north-west of Tamale. Its location near to the Nawuni dam, which is the water intake source for the GWCL-Tamale, has given it the advantage of benefiting from a public sector urban pipe-borne water supply (POI 2008), along with other communities in the Dalun-Tamale Corridor. Dalun was chosen for this study in view of its proximity to the water source and the associated water use behaviour, apart from being under the influence of the TWSP. The indigenes of Dalun see the source of water as their environmental resource, which in the context of their culture is considered as a free gift of nature (SNV 2009; Bukari 2011).

Available information indicates the Nawuni Dam is not the first intake source of water to the GWCL in the Dalun-Tamale Corridor. Before the movement to Nawuni in Dalun, the company depended on water from a dam in Wuba, north-west of Dalun in the same corridor. The over-dependence on water sources in the corridor is due to the passage of the White Volta River through this corridor, the fluvial activities of which provide adequate water for the dams throughout the year (SNV 2009). Apart from the indigenous uses of the water bodies for fishing, construction, and other domestic uses (washing, livestock rearing, cooking, etc.), the communities also sees this environmental resource as providing economic gains to the GWCL. While the treated potable water they obtain promotes their welfare through improved health, this is less well acknowledged by the community (Bacho, 2001). The environmental resource consciousness, among other factors, contributed to the emergence of conflict over water tariff determination and collection in the corridor.

Pre-TWSP water tariff determination and water-related conflicts

Water tariff determination in the Dalun-Tamale TWSP Corridor has taken various forms and procedures before and after the TWSP intervention in the area. Prior to the TWSP intervention, the determination of water tariff was solely the prerogative of the GWCL, and it was based on the lifeline rate or Increasing Block Tariff (IBT) method (Whittington 1992; Nkrumah 2004; Bukari 2011), in which initial blocks of metered water consumed attract lower tariff rates, while higher consumption levels are associated with higher tariff rates. The GWCL officials were responsible for reading the public and domestic pipe metres as well as the computation of the monthly water bills, without any form of local participation.

The District Assembly instituted multi-purpose community committees in the corridor, which before the TSP intervention were intended to serve as intermediaries between the people and any public and private sector community development actors (SNV 2009; Bukari 2011). The role of the committees in the public water services was simply to receive water bills from the GWCL for the various public standpipes and distribute these to the households making use of the facilities, as well as collecting the bills on behalf of the company. The collection of the water bills was done either at the chief's palace or by committee members going from house-to-house. The non-participatory nature of the tariff determination process, among other factors, generated water tariff-related conflict in the corridor, and in Dalun in particular.

Conflict over water tariffs emanates from a myriad of factors relating to the nature of services, and the determination and collection of tariff between public water users and the GWCL in the Dalun-Tamale TWSP corridor. The focus is on the clash of interests, the struggle to resist or overcome opposing forces or powers; leading to a state of antagonism and discord, and how these impacted on water tariff collection and payment, as well as the associated disequilibrium that results from the conflict between the indigenes of Dalun and the GWCL.

According to SNV (2009) the traditional belief about water as a free gift of nature previously influenced negative attitudes towards payment for water. This belief was also entertained prior to the structural adjustment programmes (SAP) in Ghana in the early 1980s, as the supply of public water was virtually free. The shift to cost recovery, which was re-emphasised in the TWSP regime, therefore contravened the traditional values of the indigenes, and limited their right to use the water freely as an environmental common pool resource. These notion of traditional values also appear in the findings of Zibechi (2008), and through the water war of April 2000, the poor of the city and countryside of Cochabamba (Bolivia) succeeded in expelling the multinational corporation which tried to charge them for this most basic common good (water).

On the other hand, the GWCL argues that the cost of physical infrastructural outlay, and the treatment of the water, make the potable water supplied to the people different from the ordinary water in the Nawuni Dam or the White Volta River (to which no one denies the people the right of free access and use). These issues had the potential of breeding both environmental and value-based conflicts (Botchwey 2006; Mahama 2010), which, at the emergence stage, remained non-violent.

Another cause of the emergence of conflict over water tariffs in the area was related to the nature of public water services. According to the TWOP report (2007), before the TWSP intervention in pro-poor water tariff issues, the water infrastructure was old and delivery and service pipelines could no longer withstand high pressure (see also Nkrumah 2004). This increased the number of technical breakdowns of standpipes and the attendant problem of inter-ruption of services to affected zones. However, despite the temporary termination of services, water tariffs for such standpipes, especially unmetered ones, were charged at rates equivalent to what was obtainable for normal services. In response to this, locals argued that they would not pay for water they did not consume (SNV 2009). Thus, the conflict of interest between utility maximisation and the profit motive of the water users and the suppliers respectively, was necessary for the stimulation of a non-violent interests conflict.

Additionally, there were cases of suspected manipulation of water tariffs by GWCL officials to rates higher than the actual metre values, thus implying the emergence of a data-based conflict between supplier and consumers (SNV 2009). The limitation to this claim of misrepresentation by the locals was that it was based on intuition without empirical facts. Before the TWSP, the com-munity committees of the Dalun-Tamale Corridor received no training on metre reading and interpretation for the computation of monthly water tariffs. Additionally, at that time, community participation in public water services was minimal, and so there was no regulation of the flow of the public standpipes, leading to the waste of water by women, children, herdsmen, and domestic masons (SNV 2009). This had the tendency of pushing the water being used at public standpipes to higher blocks beyond the lifeline rates provided by the IBT method, thus making the tariffs more regressive for the poor and so defeating the aim of the IBT to safeguard the interest of poor public water users (see also Whittington 1992).

Furthermore, there were also intra-societal sources of the emerging water tariff conflict, ema-nating from the clash of interests between households and community committee members over suspected misappropriation of water tariffs collected by the latter. The signs of conflict resulted when current monthly water bill statements did not reflect adequate settlement of previous bills, leaving the impression that the committee members did not pay the right amounts collected from the households to the service provider (SNV 2009; Bukari 2011). Interestingly, the potential of this issue to erupt into violent conflict is multi-dimensional. First, the committees not only owed accountability to the households for the collected water tariffs, but also to the GWCL. Second, whereas both households and the GWCL demanded accountability from the committees, the committees also accused the GWCL of failing to pay commissions the company owed them for collected water bills paid. The unanswered question therefore remains: is the failure of the committees to make full payment of collected water tariffs to the GWCL a retaliation for the non-payment of the commission the company owed them?

On the other side of this issue, the GWCL stated that there was an agreement between the company and the committees, that 20% of collected water bills rightly paid to the company was payable to the committees as commission if, and only if, the committees were able to collect 100% of the monthly water tariffs from the households and duly paid to the company (Bukari 2011). This situation was therefore a potential cause of both intra-societal and interest conflicts, which remained non-violent, but gradually approached escalation stage.

Finally, economic factors were also responsible for the emergence of the water tariff conflict in the area. According to POI (2008), about 85% of the people of Dalun are engaged in food crop farming. The rain-fed nature of this sector in northern Ghana means that the income flow of the farmers is also annual (Kendie 1992). Thus, immediately after harvest, farming households are monetised enough to adequately pay for utility services including water. By the middle of the year, especially during the cropping season (between June and early September), they run out of the means of subsistence and can hardly meet monthly bills for water and other basic needs (SNV 2009). Thus, the people of Dalun argued that given the low and seasonal nature of their household income, the water tariff rates are not only high, but also the periodic payment of the tariffs on a monthly basis is not suitable for their economic conditions. Hence, they expressed the view that a lump sum payment of water tariffs after harvest only once a year would be more favourable (Bukari 2011). On the other hand, the GWCL was of the view that the tariff determination method (IBT) was designed to suit their economic conditions, thus giving rise to the emergence of economic conflict over water tariffs.

The escalation of water tariff conflict was the stage at which there was evidence that a conflict existed, and it was characterised by negative attitudes such as destructive, rebellious, and non-conforming water use behaviours. Though the escalation of the water tariff conflict in the Dalun-Tamale Corridor was not so violent that it could claim life, it manifested itself in a number of ways including water tariff evasion, illegal tapping of water by breaking through main delivery pipelines, and households' non-cooperation with committee members and GWCL officials. These acts were manifestations of the destructive consequences of conflict and expressions of social disorganisation, which according to Byron and Robert (1989) in their social disorganisation theory, can arise when there is gross mismatch between the income status of a society and conditions of social welfare expectations (such as high utility tariffs) among others. The influence of tariff rates on economically related violent conflicts was further supported by the Criminal Justice Policy Review (CJPR) (1999), which argued that the ratio of tariff contributions to the total number of returns or value of services is positively related to violent conflict.

Growing conflict negatively impacted on collected water tariffs by the committees in the year immediately before the initiation of the TWSP model in the area (2006). Table 1 presents the total water tariffs for 2006, and the actual amounts settled by the three selected zones in Dalun, and two other communities considered to be along the corridor, which were used as a control group. These communities, though of the same ethnic background and the same socio-economic level, did not border the Nawuni Dan and did not consider it as their natural environmental resource. The first three communities in Table 1, Dalun Station, Dalun Nayili Fon, and Dalun Kanbong Naa Fon are subsections of Dalun, which collectively consider the Nawuni Dam and the White Volta River as their environmental resources. This claim on the water resource influences their water use behaviour negatively, especially in terms of payment for water (see Whittington et al. 1991; Kendie 1992).

Table 1 shows that for the service year 2006 (before the TSP intervention), the total annual water tariffs (debit balances) were GH¢1441.17, GH¢2599.91 and GH¢1675.99 for Dalun

Table 1. Water tariffs collected during the period of tariff-related conflict (2006) before the TSP in Dalun.

	Year	Total debit (GH¢)	Total credit (GH¢)	Percentage of bills settled (%)
Water-resource owning communities:				
Dalun Station	2006	1441.17	158	11
Dalun Nayili Fon	2006	2599.91	158	6
Dalun Kanbong Naa Fon	2006	1675.99	158	9
Overall percentage of annual water tariff collected	**2006**	–	–	**8**
Non-water resource owning communities:				
Gbolo	2006	415.12	420	101.2
Sorugu	2006	376.98	348	92.3
Overall percentage of annual water tariff collected	**2006**	–	–	**97**

Source: Commercial Unit, Ghana Water Company Ltd – Tamale (2006 Report).

Station, Dalun Nayili Fon, and Dalun Kanbong Naa Fon respectively. Of these amounts, the community committees collected an equal amount of only GH¢158 for each the zones. The figures represented 11, 6, and 9% of the total tariffs for the respective zones, and 8% of the overall annual water tariffs collected for the three zones. There was no precise explanation for the equal payment by the three zones, but from the findings of Bukari (2011), in view of the conflicting situations expressed earlier in this paper, the nature of the amounts paid was influenced by the collective decision of the leadership of the zonal water committees, whose standpipes involved were metered at that time.

A close look at the pattern of payment also reveals the extent of the influence of the degree of indigenousness on attitudes towards water tariff payment: in the Dagomba language, *Dalun Nayili* stands for Dalun Chief's Palace; *Dalun Kanbong Naa* for the Dalun Chief Executioner; while *Dalun Station* as the Central Business District of the area is ethnically heterogeneous. Thus, the indigenous leaders and their kinsmen exhibit more negative attitudes towards tariff payment compared to the non-royal settlement (Dalun Station).

Conversely, looking at the debit and credit columns for Gbolo and Sorugu in Table 1, which are not part of the water resource-owning communities, the tariff results were very satisfactory as they were able to settle over 100% and 92% respectively, while the overall annual water tariff collected was 97% of the total debit figure. The apparent, satisfactory performance of Gbolo and Sorugu may be assumed as due to the absence of the influence of environmental resource ownership on their water use behaviour. Additionally, the absence of considerable natural water bodies meant that these communities had no alternative other than to pay adequately for the quality potable water services of the GWCL if they wished to sustain the services, thus reducing the tendency of tariff-related conflicts between the people and the company to escalate. The ensuing sections present the actual nature of the tariff conflict and how the TWSP model was used to address the conflicts through indigenous and modern methods.

The TWSP water tariff determination and conflict resolution approach

After the formation of the TWSP in Dalun, community participation was incorporated into the tariff determination process through formally instituted Community Water Boards. The boards were chaired by the traditional chief of each TWSP pilot community in the corridor (SNV 2009; Bukari 2011). The water boards were given some training on facility maintenance, metre reading, tariff collection, record keeping, and other aspects that built upon their capacities to effectively participate in water-related decision-making and to reduce water tariff conflict.

In view of the fact that the major objective of forming the TWSP was to resolve the water tariff conflict that existed between the GWCL and the indigenes, which necessitated the drawing in of the private sector (SNV and POI for Dalun in particular), the TWSP made use of mediation, since it involved the use of the third party (private sector developers). The purpose was to facilitate better communication and respect for each other's ideas in the collective management process by entering into legal relationships and clarifying role definitions, as discussed earlier. This approach was both modern and traditional as the key players involved formal sector organisational representatives from the GWCL, SNV, and POI, as well as indigenous structures including the traditional chiefs (such as the *Dalun Naa*) and the Community Water Boards made up of the indigenes. This approach generated a series of other forms of resolution of specific conflicts relating to the tariff situation.

Initially, as the environmental resource conflict over ownership of water sources was also related to the traditional values of the people, the TWSP realised that the best approach was to use the indigenous structures such as the traditional chief (*Dalun Naa*) and the Community

Water Board members. Through the use of intra community conferences or community forums organised by the local structures, households of the various communities were educated on the difference between natural water from the river or dam and potable water from standpipes, in order to debase the traditional notion of the free gift of nature, in view of the cost differentials. Those whose attitudes towards tariff payment remained negative, were dealt with by the traditional adjudication, in which the Community Water Boards reported deviants to the *Dalun Naa* and his council of elders as the traditional court of law. The traditional court could then expedite the verdict, by setting deadlines for settlement of the tariff or stopping the defaulting households from using public water until payment was made. The order is usually executed by the Water Board and enforced by the Chief Executioner (The *Dalun Kanbong Naa*). The same approach is used for the settlement of reported cases of illegal tapping of water. In this case, special punishment was meted out based on the collective decisions of the traditional court and the GWCL.

The conflict over tariffs without water services due to unserviceable public standpipes was also resolved through negotiations for cost-sharing in the maintenance of water infrastructure. The negotiation between the Zonal Water Boards of Dalun (chaired by the Chief of Dalun) and the GWCL, which was facilitated by SNV in 2007, led to the adoption of a community ownership approach in the water system maintenance. The community became responsible for the maintenance of service components (such as service pipelines and standpipe components), as well as reporting any other technical faults to the company as soon as they were detected. It was also provided that each Water Board should have a technician, such as a plumber, trained by the GWCL to attend to minor breakdowns on the service pipelines and other components. The GWCL was to be responsible for repairs on delivery or supply mains (such as intake and delivery pipelines), as well as general rehabilitation of the water system (SNV 2009; Bukari 2011).This combination of action through negotiation between formal and indigenous structures reduced instances where standpipes faults persisted for long after bills continued to flow.

The data conflict, resulting from mistrust in the validity of metre readings and the associated high tariffs, was also resolved through mediation (Schellenberg 1996), in which the private sector partners of the TWSP (SNV and POI), entreated the GWCL to train and involve the Water Board members in metre reading and the computation of the monthly water tariffs for each standpipe, and to require Water Board attendance as witnesses whenever metres were read (POI 2008; SNV 2009; Bukari 2011). Using this approach, the indigenes could determine their monthly water tariffs in advance, and so prevent the tariff data conflict.

The intra-societal conflict resulting from the misappropriation of collected water bills by the former community committee members was resolved by the avoidance approach, specifically by the diplomatic withdrawal (see Bhaskaran 2003) of the community committees from handling community public water issues. This was formalised by the replacement of the committees with the Community Water Boards, specially trained to represent the locals in community public water-related issues, including water tariff collection (Bukari 2011). The Water Boards were trained to be more accountable to the households and the GWCL, through proper bookkeeping and financial reporting formats (Bukari 2011). This increased the confidence of households to pay their tariffs to the board members, and improved the relationship between the boards and the company for better services.

For the economic conflict resulting from the small and seasonal nature of household income, *collective bargaining* between the Water Boards and the GWCL under the chair of the *Dalun Naa* proved workable. There was a shift from the IBT to bargaining for the minimum tariff rate. This led to the adoption of a minimum chargeable consumption level of 18 litres at a tariff rate of 2Gp for the year 2007. Households also expressed the view that given the seasonal nature of their income flow, water tariff should be charged after the harvest period, since they become short of income at certain parts of the year (pre-harvest), but cannot do without potable water. This was a call for agreement to adjust the terms of tariff payments between the GWCL and the service consumers by making considerations for the economic conditions of the latter, which is a form of *compromise*. The failure of the GWCL to accept this condition reflected a *win-loss* conflict resolution outcome, as the interest of one group was sustained while that of the other was denied.

The TWSP resolution of water-related conflicts, and tariff collection and payment

Table 2 presents the data on debit and credit balances of water tariffs for 2008, after the conflict resolution processes under the TWSP model.

The data in Table 2 reflects the positive effects of the indigenous and modern conflict resolution methods adopted by the TWSP model in Dalun. It shows that collected water tariffs after the intervention improved for all the three zones of Dalun, as the percentages of annual water tariffs settled improved from 11, 6 and 9% (for 2006), to 32.5, 8 and 27% for Dalun Station, Dalun Nayili Fon, and Dalun Kanbong Naa Fon respectively for 2008. Table 1 revealed that the Community Committee in the midst of the tariff conflicts in 2006 collected only 8% of the total annual water tariff in Dalun, while while Table 2 shows that the new Water Boards collected up to 21% of the tariffs in 2008 (see also Bukari and Abagre 2013). The non-water resource owning communities, on the other hand, dropped behind the TWSP communities as they paid only 20% of the total annual water tariff.

The tariff figures for 2008 (after the TWSP intervention) for the resource and non-resource owning communities reflect a similar pattern of distribution of debit figures, and the favourable credit balances for the former. The improvement in tariff payments in Dalun communities at the same time as the payments dropped substantially in the non-resource owning control provides evidence to suggest that the TWSP was effective in improving water tariff payments and resolving disputes through traditional and modern conflict resolution. Of course we need to track changes in water tariff payments over time to assess trajectory and sustainability, but this case study suggests a basis for follow-up research. Bukari's findings indicated that there was a statistical significance of 0.006 at a contingency coefficient of 0.48 for the relationship between the role of the TWSP in conflict resolution and willingness to pay for water (Bukari 2011, 148). The significance level was below the cut-off point of 0.05, which is the requirement for the acceptance of an alternative hypothesis that a relationship exists between two or more variables, and the strength of this

Table 2.   Water tariffs collected after conflict resolution by the TSP in Dalun.

	Year	Total debit (GH¢)	Total credit (GH¢)	Percentage of bills settled (%)
Water resource owning communities:				
Dalun Station	2008	1617	525	32.5
Dalun Nayili Fon	2008	2180.75	166	8
Dalun Kanbong Naa Fon	2008	1959.92	536.6	27
Overall percentage of annual water tariff collected	**2008**	–	–	**21**
Non-water resource owning communities:				
Gbolo	2008	2268.61	446	38.6
Sorugu	2008	1716	355	20.1
Overall percentage of annual water tariff collected	**2008**	–	–	**20**

Source: Commercial Unit, Ghana Water Company Ltd – Tamale (2008 Report).

relationship is measured by the value of the coefficient of correlation (Sirkin 1999), which was 0.48 or 48% in the case of Dalun.

In the non-water resource owning communities, the opposite was the case after the TWSP intervention in Dalun. Gbolo and Sorugu, which fall in this category, were not part of the TWSP pilot communities in the Dalun-Tamale Corridor, hence their tariff performances in public water services were not influenced by the TWSP. Table 2 shows that while the Dalun zones improved upon tariff settlements, Gbolo and Sorugu had surprising declines in the annual tariffs settled, reduced to 38.6 and 20.1% in 2008 from 101.2 and 92.3% respectively in 2006. Bukari's (2011) focus group discussions with standpipe water users in Gbolo and Sorugu indicated that the poor tariff payment record was due to the excessively high increase in water tariffs within the period (see data in Tables 1 and 2). On the issue of why the tariffs increased within these areas, interview results with GWCL officials yielded two reasons; the first was due to the high cost of the water infrastructure rehabilitation under the TWOP within the period. The second reason was that most standpipes that were previously not metered before the TWOP were now fitted with metres to reflect the actual consumption levels of users. Since they share similar economic values with the Dalun community (the same ethnic group [Dagomba], rural and agrarian), they became affected by the environmental and economic conditions that negatively impacted on Dalun in water tariff issues before the TSP, thus demanding a similar intervention in these areas.

Though arising long after the TWSP intervention between 2007 and 2008, there is an ongoing debate as to whether or not to switch to the use of prepaid metres in public water services in Ghana. The issue is arising because of the high level of water tariff evasion, unaccounted-for water, and increasing tariff arrears. This is an indication that challenges of water tariff payment are neither limited only to the Dalun community nor to rural areas, but a national issue. In every case, however, the special needs of the poor should be considered, since water is a basic necessity of life (Bacho 2001).

Discussion and conclusion

The results have shown that the local people epistemologically perceived the water source for the Ghana Water Company Ltd. as a common pool environmental resource, which the community felt was free but from which the GWCL received economic gains through tariff imposition on services. The contradiction between the tariff imposition and the indigenous metaphysical position regarding water as a free gift of God negatively impacted on the community's willingness to pay. Thus, the emergence of water tariff conflict between the service provider and the indigenes appeared to have been justified on the grounds of unequal interests in cost recovery and social welfare maximisation between the two parties respectively. It was, however, obvious that the absence of authentic and traditional community participation before the identification of the need for the tri-water sector partnership accounted for the unexpected results in water tariff collection and payments.

The establishment of the TWSP in Dalun not only demonstrated the strength of partnership in addressing community development-related conflicts, but also provided a practical example of the positive relationship between indigenous and modern conflict resolution methods in addressing the water tariff conflict in particular. The results indicate that generally, approaches used in indigenous and modern conflict resolutions are similar, but the major differences lie in the structures or key players. Thus, whereas indigenous approaches involve the traditional political set-up, the elders, the youth (in the Water Boards), and households, the modern system involves formal sector structures, such as public and private sector organisations.

The apparent improvement of the water tariff payments after TWSP intervention in Dalun is a manifestation that indigenous conflict resolutions methods, combined with modern methods, are best suited for managing escalating water conflicts. Neither indigenous nor modern conflict resolution methods on their own are best suited for dealing with the water tariff conflict that escalated; best results are achieved by integrating both approaches. The TWSP was effective in the water tariff conflict resolution, and results have empirically proven that successful TWSPs can ensure that the public sector will serve its constituents better, businesses will profit, and communities will benefit.

Recommendations

Given that the methods used in the conflict resolution process did not reflect typical Dagomba ontology, such as spiritually, it is recommended that methods such as sacrifices to ancestral spirits for peace and cooperation between the people and external development agencies; and the use of appropriate traditional communication forms such as music and dance, among others, could be adopted as preliminary stages before other traditional and modern integrated forms of conflict resolution are implemented by the formal structures that constitute the partnership. This ensures that the basis of conflict and the need for change are well communicated at the grassroots level, which is essential for the understanding and cooperation of those at the grassroots level with decision makers towards consensus building.

Furthermore, given the poverty situation, the partnership activities could also be extended to address issues of capacity building for economic empowerment. Without an increased ability to pay, it is doubtful whether partnership interventions for conflict resolution alone would be adequate for meeting public sector targets for cost recovery in pro-poor water services. Moreover, the current debate on the transition to prepaid metres for public water tariff determination in Ghana could face serious challenges given the present state of poverty in the area, unless something is done to increase the ability of women, in particular, to pay for their water requirements on sustainable basis.

Note

1. There is a substantial literature on the concepts in the ongoing dialogue around endogenous approaches to conflict resolution, and about water conflicts in particular. The authors have written separately on these issues, but the length of this article does not permit an exploration of key concepts. This article instead attempts to provide a useful case study of the integration of traditional and modern conflict resolutions approaches in one instance of water tariff conflicts.

References

Bacho, F. Z. L. 2001. *Infrastructure Delivery under Increasing Poverty: Potable Water Delivery through Collective Action in Northern Ghana.* Baroper: Spring Center – University of Dortmund.

Bhaskaran, W. M. 2003. "Role of Academics in Conflict Resolution." Accessed October 16, 2010. www.mkgandhi.org/nonviolence/academics.htm

Botchwey, G. A. 2006. *Steps to Self-Reliance: For Groups and Communities*. Cape-Coast, Ghana: Catholic Mission Press.

BPD (Business Partnership for Development). 2008. "International Development." Accessed September 15, 2010. http://www.en/wiki/wikiProject

Bukari, F. I. M. 2011. "An Assessment of the Tri-sector Partnership Model in Pro-poor Water Tariff Collection at Dalun, Northern Region, Ghana." Masters thesis, University for Development Studies, Ghana.

Bukari, F. I. M., and C. I. Abagre. 2013. "An Alternative Model for Pro-Poor Water Services: Improving Water Tariff Payments in Low Income Communities." *Journal of Environment and Earth Science* 3 (8): 157–165.

CJPR (Criminal Justice Policy Review). 1999. "Social Altruism: Tax Policy Review: A Cautionary Tale." *Sage Publications*, 10 (3).

Commercial Unit, Ghana Water Company Ltd. 2006. Annual Statement of Water Bill. Tamale: GWCL.

Commercial Unit, Ghana Water Company Ltd. 2008. Annual Statement of Water Bill. Tamale: GWCL.

Corbin, J., and Strauss, A. (2008). *Basics of Qualitative Research: Techniques to Developing Grounded Theory* (3rd Ed.). Los Angeles, CA: Sage.

Coser, L. 1956. *The Functions of Social Conflict*. Philadelphia: The Free Press.

Fink, C. F. 1968. "Some Conceptual Difficulties in the Theory of Social Conflicts." *Journal of Conflict Resolution* 12: 412–460.

Kane, E. 1995. *Seeing for yourself: Research Handbook for Girls' Education in Africa*. Washington D.C: World Bank.

Kendie, S. B. 1992. "Survey of Water use Behaviour in Rural North Ghana." *Natural Resource Forum* 16 (2): 126–131.

Kumar, R. 1999. *Research Methodology*. New Delhi: Sage Publications.

Mahama, E. S. 2010. "Conflict in Ghana: Strategies, Stakeholders and the Way Forward." In *Conflict Management and Peace Building for Poverty Reduction*, edited by B. S. Kendie, University for Development Studies, 130–158. Tamale-Ghana: Center for Continuing Education and Inter-disciplinary Research.

Nkrumah, M. K. 2004. "Challenges to Potable Water Management in Ghana: Making Public- Private Partnership Work." *Ghana Journal of Development Studies* 1 (2): 95–100.

POI (Pragmatic Outcomes Incorporated). 2008. *Facility Management Plan for Dalun Water Boards*. Tolon: Tolon-Kumbungu District Assembly.

Schellenberg, A. J. 1996. *Conflict Resolution, Research, Theory and Practice*. Albany: State University of New York Press.

Sirkin, M. R. 1999. *Statistics for Social Sciences* (2nd ed). Thousand Oaks, California: Sage Publications.

SNV (Netherlands Development Organization). 2008. "Annual Report on the Activities of the Tri-sector Partners." Unpublished.

SNV (Netherlands Development Organization). 2009. "An Assessment of the Tri-sector Partnership Participatory Model of Pro-poor Water Delivery Services in the Dalun-Tamale Corridor." Unpublished.

Twumasi, P.A. 2001. *Social Research in Rural Communities*. Accra: Ghana Universities Press.

Whittington, D. C. 1992. "Possible Adverse Effects of Increasing Block Water Tariffs in Developing Countries." *Economic Development and Cultural Change* 41: 75–78.

Whittington, D., T. Donald, D. Lauria, and X. Mu. 1991. "A Study of Water Vending and Willingness to Pay for Water in Onitsia, Negeria." *World Development* 11 (2–3).

Zibechi, R. 2008. "*From Water War to Water Management: Americas Program*." Accessed February 16, 2010. http://ircamericas.org/esp/6130

Endogenous African governance systems: what roles do women play in rural Malawi?

Chimwemwe A.P.S. Msukwa and Marion Keim-Lees

Endogenous African governance systems are criticised for excluding women. This critique ignores several realities that women have played roles different from those of men. This article examines the roles that women play in endogenous governance structures of patrilineal and matrilineal ethnic groups in rural areas in Malawi on leadership, violent conflict prevention, and transformation. It argues that these endogenous governance systems inherently contain features that enable women to actively participate and play powerful leadership roles, though men dominate in terms of numbers and authority. These gender patterns do not seem to change much despite the changing political, social, and economic environment.

Les systèmes endogènes africains de gouvernance sont critiqués parce qu'ils excluent les femmes. Cette critique ignore la réalité selon laquelle les femmes ont joué des rôles différents de ceux des hommes. Cet article examine les rôles que jouent les femmes dans les structures de gouvernance endogènes des groupes ethniques patrilinéaires et matrilinéaires dans des zones rurales du Malawi sur le plan du leadership, de la prévention des conflits violents et de la transformation. Il soutient que ces systèmes de gouvernance endogènes présentent de manière inhérente des caractéristiques qui permettent aux femmes de participer activement et de jouer de puissants rôles de leadership, même si les hommes dominent en termes de nombre et d'autorité. Ces schémas de genre ne semblent pas beaucoup changer malgré l'environnement politique, social et économique en mutation.

Los sistemas endógenos de gobernanza en África han recibido críticas por el hecho de excluir a las mujeres. Sin embargo, tales críticas pasan por alto la realidad de que las mujeres han desempeñado roles diferentes de los cumplidos por los hombres. El presente artículo examina los roles que, en las áreas rurales de Malaui, han desempeñado las mujeres en los ámbitos de liderazgo, de prevención de conflictos violentos y de transformación, dentro de las estructuras de gobierno endógenas de los grupos étnicos patrilineales y matrilineales. Al respecto, se sostiene que, aun cuando los hombres dominen en términos de presencia numérica y autoridad, dichos sistemas contienen modalidades inherentes que permiten la participación activa de las mujeres y que éstas desempeñen roles fuertes de liderazgo. Estos patrones de género no parecen haberse alterado mucho a pesar de los cambios ocurridos en los ámbitos políticos, sociales y económicos.

Introduction

One of the criticisms of traditional African peace and justice systems has been their tendency to exclude women. It has been argued that some African traditions have not always promoted gender equality (Murithi 2006). This critique ignores several realities. Historically women played roles different from those of men. Sometimes they were excluded from power and decision-making outside the household. In other cases women held respected and powerful roles in economic and political decision-making and conflict resolution. When social change occurs, it is because societies build upon their traditions and make adjustments to adapt to the changing environment.

In Africa, we can look at historical gender relations and see multiple versions of exclusion and inclusion of women. In the pre-colonial traditional societies of Rwanda for example, while women participated as part of the face-to-face communities attending the old *gacaca*[1] sessions, men dominated (Ingelaere 2008). In the *Pokot* pastoral communities in Kenya, apparently the roles of women were not as prominent as those of men. It was mostly senior elderly women who contributed to proceedings in a *kokwo*.[2] The roles of other women mostly included documenting the outcomes of the *kokwo* for reference in future meetings; providing advice to the council on what to do and what not to do, citing prior occurrence or cultural beliefs; and voicing their opinions when asked to do so (Pkalya, Adan, and Masinde 2004). Within the Issa and Gurgura communities in Somalia women are not allowed to participate in formal conflict resolution mechanisms due to the patriarchal nature of these communities (Tadesse, Tesfaye, and Beyene, 2010).

In Malawi, we can see multiple patterns of gender roles in matrilineal and patrilineal societies. This article examines the roles that women play in the traditional socio-political governance structures of the patrilineal ethnic groups of the Sukwa and Ngoni of Mzimba district, as well as in matrilineal ethnic groups of the Chewa and Yao in rural areas in Malawi with respect to leadership, violent conflict prevention, and transformation. It is based on the findings of field research conducted by the principal author between 2008 and 2010 in the four ethnic groups. Critical examination of these roles shows that traditional governance systems of matrilineal and patrilineal societies in Malawi do not only exclude or undervalue women as often assumed, but they also inherently contain features that enable women to actively participate as well as play powerful leadership roles. This implies that development approaches targeting these societies should be designed in ways that build on the existing strengths in gender relations if they are to effectively address the needs of these societies.

Case study context

The study was conducted in Malawi, a small country in southern Africa that shares boundaries with Tanzania to the North, Mozambique to the South and East, and Zambia to the West.[3] The country is predominantly rural, with about 90% of the population living in rural areas under traditional ethnic leadership. Malawi has 11 ethnic groups mostly distinguishable by languages, lineage, and customary practices. On the basis of lineage, Malawi ethnic groups can be broadly divided into two: (1) matrilineal ethnic groups, which trace their lineage from the mother – these include the Chewa, Yao, and Lomwe ethnic groups; and (2) patrilineal ethnic groups, which trace their lineage from the father – these include the Ngoni, Tumbuka, Tonga, Sena, Ngonde, Lambya, Sukwa, and Nyika.

Each of these ethnic groups has an elaborate socio-political governance structure. The basic framework of these socio-political governance structures is similar across both patrilineal and matrilineal ethnic groups. The structure comprises individuals (men, women, and children) as the primary building blocks; family units, which include the nucleus families and the extended families; as well as the traditional leadership unit which links individuals from different family units as well as different families. The traditional leadership structure has a hierarchy, which is common to all the ethnic groups and comprises the following in order of increasing authority – the village, the group village, and the traditional authority. The village comprises several nucleus or extended families within a certain demarcated geographical boundary under the leadership of a village head person. A group village comprises several villages (often three to 10) under the leadership of a group village head person.

The role of women in the socio-political governance structures

Regardless of whether the ethnic group is patrilineal or matrilineal, women form an integral part of the socio-political governance structures of the Sukwa, Ngoni, Chewa, and Yao tribes in Malawi, particularly in their individual capacities as members of nucleus and extended families as well as the village community. Women also play some leadership as well as other prominent roles in the socio-political governance structures of both matrilineal and patrilineal ethnic groups.

Within the patrilineal tribes, a woman may be installed as a village head or group village head in very exceptional cases particularly where there is no man in the "royal family" to take over the position. During the time of the study, all the 52 villages in the Sukwa ethnic group were headed by men except one. Within the Ngoni communities, there were no females holding any traditional leadership position.

There are however more women holding traditional leadership positions in the matrilineal ethnic groups. In the Chewa ethnic group, for instance, it is common to see a woman holding any of the traditional leadership positions – village head, group village head, and even the position of traditional authority. Though fewer than in the Chewa ethnic group, there are significant numbers of women holding traditional leadership positions of village head person and group village head person in the Yao ethnic group. Men dominate in the traditional leadership of the latter, probably because of the influence of Islam, the predominant religion in this ethnic group.

Women also play some prominent roles within the extended families of both matrilineal and patrilineal ethnic groups. In the Ngoni, Chewa, and Yao ethnic groups, elderly women play the role of conveners of special forums for women and girls. These forums create platforms for women and girls in a particular extended family to chat, share experiences and jokes, and learn from each other as well as resolve differences. In addition, women also use the forums to teach the girls the values, beliefs, and principles of their culture.

For matrilineal ethnic groups, the *mbumba* (female elders in the chiefs' extended families) play a very important role of selecting the new traditional leaders and their *nduna* (advisors of the traditional leaders). In both the Chewa and Yao matrilineal ethnic groups, the village head person, group village head person, and traditional authority are all selected by female elders of their extended families.

In the Ngoni ethnic group, the extended family selects four elderly women, two from the husband's side and two from the wife's side, to advise the newly married husband and wife on how they should live together. In the Sukwa ethnic group the wife of *ifumu*, the leader of a sub-village comprising several extended families, is respected and plays the role of her husband in his absence.

Although there are relatively more women holding traditional leadership positions in matrilineal ethnic groups of the Chewa and Yao, men still dominate in numbers. Similarly, though overall in both matrilineal and patrilineal ethnic groups, women play some leadership roles within traditional leadership structures and the extended families, and their authority and performance are just as good as those of men, still men dominate in traditional leadership roles.

The role of women in prevention of violent conflict

Women play important roles in the prevention of violent conflicts within their families and communities in both patrilineal and matrilineal ethnic groups. These roles mostly relate to socialisation of children – women play the crucial role of moulding their children, particularly the girls, to enable them to acquire and internalise the moral values of their society. They teach, continuously advise and monitor the behaviour of children, ensuring that they have adopted the necessary moral values that will enable them to live in harmony with others in the family and the wider community.

Patrilineal and matrilineal ethnic groups in Malawi share common moral values, which influence the way individuals related to each other. These moral values include respect for each other, maintaining relations and relationships, interdependence, unity, kindness, friendliness, sharing, tolerance, self-restraint, humility, love, obedience, trustworthiness, and transparency. These moral values prevent individuals from displaying or engaging in aggressive behaviour but rather enhance self-control or self-restraint and promote relationships. For example, mutual respect between men, women, and children and for parents, elders, and leaders helps to prevent violent conflict. Respect enables individuals to value other people including their contributions and ideas. It restrains individuals from displaying aggressive behaviour towards others. It enables one to seek more tactful ways of engaging with those one respects and in doing so, averts potentially violent conflict situations.

In both patrilineal and matrilineal ethnic groups, women have the responsibility of passing on moral values to girls and younger boys as they spend more time with them compared to men. They play the primary role of bringing up children, taking care of not only their physical and biological needs, but also socialising them. In this respect, the women do not only assist their biological children, but also any other girls and boys around them within the community. The socialisation process enables the children to internalise the moral values which make them live in harmony with others.

In the matrilineal ethnic groups of Chewa and Yao, the socialisation process includes initiation ceremonies for girls and boys. There are special female elders of the community, locally termed *nankungwi* (an elderly woman with deep knowledge of culture who is chosen by the community to be in charge of initiation of girls), who play prominent roles in the initiation ceremonies to infuse moral values in girls. They confine the girls in special camps where they teach them different cultural values and practices. In the Chewa ethnic group the initiation ceremony for girls (*chinamwali*), is to a great extent purely a women's affair, as the key roles of *nankungwi* and *mphungu* (tutor) are played by women (Van Breugel 2001, 190)

The role of women in conflict transformation

Women play different roles in conflict transformation processes within the nucleus family, extended family, and village community.

First, as mothers in the nucleus families, in all the four ethnic groups, women take leading roles in facilitating discussions and providing advice when violent conflicts arise between their children. Second, women are part of the various forums that discuss violent conflict at the different levels of the socio-political governance structures of both the patrilineal ethnic groups of Sukwa and Ngoni and matrilineal ethnic groups of Chewa and Yao. Within these forums, women play different roles in violent conflict transformation processes. There are however minor variations in the roles played by women across the four ethnic groups, not necessarily split between patrilineal and matrilineal ethnic groups. In the Sukwa ethnic group, at the extended family, sub-village, and village forums, women actively participate by asking probing questions, cross-examining the disputants, and bearing witness and taking part in decision-making processes alongside men as part of *bachisu* (owners of the land). The wife of *ifumu* (leader of sub village to whom the village headman delegates authority to convene and facilitate forums to discuss violent conflicts arising between members of the sub village) convenes the sub village forum to discuss a violent conflict that has arisen between two people in the absence of her husband. However, she has to invite a male *ifumu* from another sub-village to facilitate the conflict discussions.

In the Ngoni ethnic group, some forums (this can be at village, group village, or traditional authority level) have elderly women as part of the elders facilitating the conflict transformation processes.[4] The women in the facilitation teams are actively involved in cross-examining the conflicting parties and are fully involved in all the decision-making processes in the forum. In addition, within the Ngoni culture, elderly women, aunts, and grandmothers play crucial roles in domestic violent conflicts. When a dispute arises between a husband and wife, the wife first approaches the husband's aunt (sister of the husband's father) or husband's grandmother who facilitates discussions between the two. If the aunt and grandmother fail to resolve the dispute between the husband and wife, four elderly women, two from the husband's side and two from the wife's side, come in to facilitate a discussion to resolve the dispute as well as advise the couple on how they should live together. When a dispute arises between women from the same extended family, there are special female elders of the extended family who convene a *sangweni* (forum) for female members of the extended family to discuss the dispute. The elderly women facilitate the conflict transformation process in this forum.

In the Chewa and Yao ethnic groups, within the extended families, sometimes women play very active roles in facilitating forums to transform violent conflict involving both men and women, particularly where there are no competent men to facilitate the discussions. Just like in the Ngoni ethnic group, where the conflicts exclusively involve women, there are elderly women in the extended family and sometimes in the village community who convene *bwalo* (forum for discussion) for women only and facilitate discussions to transform violent conflicts that have arisen. The female elders also try to reconcile and advise the women in these forums on how best to relate to one another.

In the Sukwa, Chewa, and Yao ethnic groups when a woman holds the position of village head or group village head or even traditional authority, she automatically becomes the ultimate convener and decision maker, and assumes the overall leadership of the forums convened to transform specific conflicts. Whether these positions are held by men or women all the decisions are based on a consensus reached by the forum participants in the case of the Sukwa ethnic group, or the forum facilitation team in the cases of Ngoni, Chewa, and Yao ethnic groups.

Endogenous gender relations in the changing environment

The discussion above highlights that women are a prominent part of the traditional socio-political governance structures. They play significant leadership roles within the nucleus and extended family as well as the traditional governance structures. Women also play significant roles in the prevention of violent conflict as well as intervening as participants, conveners, and facilitators of violent conflict transformation forums. However, in general terms, culturally within the Sukwa, Ngoni, Chewa, and Yao ethnic groups, men dominate over women in both numbers and authority with respect to leadership within the traditional socio-political governance structures and the roles they play in violent conflict prevention and transformation processes.

Apparently the endogenous gender relations and patterns in the governance systems of the rural matrilineal and patrilineal societies in Malawi have not changed much over time, not only as they relate to conflict prevention and transformation systems discussed above, but also as they relate to the overall governance of society. This is despite the intensive gender campaigns associated with different development projects implemented in these societies, as well as the transformation processes that the societies are undergoing economically, politically, socially, in terms of religious beliefs and advancements in access to information, education, and communication systems.

In politics, for example, Malawian society has transitioned from pre-colonial traditional governance systems to colonial governance, to one party autocratic rule, to multiparty politics, and most recently to having a female president. In each of these political systems women have participated in political affairs in different ways as citizens. During the one party system for example, there were several initiatives to raise the profile of women in the family and society as a whole. Women were promoted through their wide involvement in traditional dances during political functions. As Kamuzu's *mbumba* (part of the president's royal family female elders) no man would stop his wife from participating in political functions, so men had less power over the political affairs of their wives. To the contrary, while one-party politics seemed to have been promoting the profile of women, women continued to play subordinate roles in the overall political governance and leadership systems. The gender relations and patterns across these political systems were not much different from those of the colonial and the traditional governance regimes.

During the era of multiparty politics, we have heard of the 50 – 50 campaigns aimed at having equal numbers of women and men in political governance and leadership structures. While there has been an increasing trend in terms of the number of women participating in political governance leadership structures, men by far continue to dominate in political leadership positions. Worse still the current mindsets of the majority of both men and women seem to support the subordinate position of women in political leadership. This implies that the changes in political systems have not contributed significantly to any shifts in endogenous gender relations and patterns.

Socially, the country has seen an increased flow of young people, particularly men, from rural to urban areas, and in some districts a significant proportion of young men have migrated to South Africa. These movements are mainly in a search for employment. In addition, HIV and AIDS have also contributed to the deaths and incapacitation of a significant proportion of the economically active groups. These factors have contributed to significant changes in the structures of the family as well as the wider society. There have been increasing numbers of not only female-headed households, but also female-dominated villages. Consequently, more women find themselves filling the leadership gaps left by men. In both matrilineal and patrilineal societies families and societies are emerging where women are beginning to dominate in leadership roles not only in the nucleus and extended families but also in the traditional leadership structures. This is more so in the matrilineal societies, where traditionally women have more influence than in the patrilineal societies. In addition, the burden on women to provide care and support for vulnerable groups, particularly orphaned children, young children in general, and the sick, as well as the aged, has significantly increased. This has limited the participation and the role of such women in other spheres of life such politics and economic activities.

Furthermore, as more and more people have migrated from rural to urban areas, in the peri-urban areas governance structures have emerged which mirror the traditional leadership structures and systems. While in the rural areas there are villages headed by village head persons, in the peri-urban areas there are blocks headed by block leaders, which have emerged to govern the people living in these areas. They deal with funeral ceremonies, conflicts, and any other social issues arising in these communities.

Each block has a male and female block leader. These have equal powers and work together in close collaboration when handling conflict, funerals, and any other social issues affecting people in their blocks. The male and female block leaders make decisions by consensus. Block leaders are elected by the majority of the people in their blocks. For a man or woman to be elected as a block leader they have to have the following characteristics – trustworthiness, should be authoritative, should command respect of the block members, must have stayed long enough to know the dynamics of the people in the block, and must demonstrate mediation/facilitation skills. A female block leader commands respect and exercises powers over the rest of the block members in the same way as a male block leader does. The leadership systems in blocks have been developed based on the traditional leadership structures. People who live in peri-urban areas often come from different ethnic groups with different prototypes of traditional governance systems. The governance systems and gender relations that have emerged in these areas have been developed through discussions, negotiations, and consensus by individuals from different backgrounds seeking to develop a common platform for their joint engagement and mutual benefit. As these negotiations have been mostly at individual level to satisfy individual interests, the governance systems that have emerged and their associated gender relations have been inclined more towards equality of men and women in leadership positions. Blocks in peri-urban areas appear to have developed new hybrid governance and leadership institutions, drawing from endogenous governance systems and gender relations as well as contemporary gender and democratic principles.

Conclusion

Women form an integral part of, and play significant roles within, the nucleus and extended family as well as the traditional socio-political governance structures of matrilineal and patrilineal societies in Malawi. Within these endogenous governance systems, women also sometimes play significant leadership roles. Men, however, take an upper hand in leadership roles in terms of numbers and authority. Overall these gender relations and patterns do not seem to change much in rural settings despite the changing environment in terms of politics, and social and economic factors. Where changes in gender relations have occurred, the societies have adjusted whatever they are already doing based on the prevailing situations as well as incorporating some contemporary principles. This implies that for any development approach to influence changes in endogenous gender relations and patterns, such an approach should unavoidably be based on a deeper understanding of the prevailing gender relations and seek ways to build on their strengths.

Notes

1. The *gacaca* means "justice on the grass" – an assembly convened in the event of a conflict within traditional Rwandan communities, where the disputing parties would be heard and judged by a panel of elders called *inyangamugayo* in the presence of family members and the rest of the community members.
2. The *Kokwo* was the highest council of elders for the *Pokot*, which also serves as the supreme court in the land. It comprises respected wise old men who are knowledgeable in community affairs and history, and good orators and eloquent public speakers who use proverbs and wisdom phrases to convince the meeting or the conflicting parties to reach a truce.
3. The study was a doctoral thesis titled "Traditional African Conflict Prevention and Transformation Methods: Case Studies of Sukwa, Ngoni, Chewa and Yao Tribes in Malawi". The research was conducted by C.A.P.S. Msukwa supervised by Professor Marion Keim at the University of Western Cape in South Africa. The research examined the traditional governance systems of the four tribes to

see if there were common cultural elements for preventing and transforming violent conflict in selected patrilineal and matrilineal tribes in Malawi, as well as selected societies from other parts of Africa. The research also examined the role of women in these endogenous systems for violent conflict prevention and transformation.

4. Previously females were not allowed to be part of the facilitation team but now they are due to of a campaign by a primary justice project implemented by the Catholic Commission for Justice and Peace in Malawi.

References

Ingelaere, B. 2008. "The Gacaca Courts in Rwanda." In *Traditional Justice and Reconciliation after Violent Conflict: Learning from African Experiences*, edited by L. Huyse and M. Salter, 52. Stockholm: International Institute for Democracy and Electoral Assistance.

Murithi, T. 2006. "Practical Peacemaking Wisdom from Africa: Reflections on Ubuntu." *Journal of Pan African Studies* 1 (4): 25–34.

Pkalya, R., M. Adan, and I. Masinde. 2004. "Indigenous Democracy: Traditional Conflict Resolution Mechanisms: Pokot, Turkana, Samburu and Marakwet." In *Intermediate Technology Development Group-Eastern Africa*. Accessed October 3, 2010. http://practicalaction.org/docs/region_east_africa/indigenous_democracy.pdf

Tadesse, B., Y. Tesfaye, and F. Beyene. 2010. "Women in Conflict and Indigenous Conflict Resolution among the Issa and Gurgura Clans of Somali in Eastern Ethiopia." *African Journal on Conflict Resolution* 10 (1): 85–110.

Van Breugel, J. W. M. 2001. *Chewa Traditional Religion*. Blantyre: CLAIM.

Putting endogenous development into practice

Nathalie Tinguery

The gap between theories and the actual practice of development is often great, but the gap between concepts of endogenous approaches and the practice of endogenous development may be hardest to bridge, particularly when the funding agency is a global actor. Nathalie Tinguery, Country Program Coordinator for US African Development Foundation (USADF) in Burkina Faso, reflects on her experience of incorporating values and goals into her development practice of working with communities and for an international funder. She describes how she remains focused on endogenous development, what this means in her development practice, and what it is about USADF policies and practice that make this brand of endogenous development possible. Views expressed are those of the author and do not necessarily represent the official positions of USADF.

L'écart entre les théories et les pratiques réelles du développement est souvent important, mais l'écart entre les concepts d'approches endogènes et la pratique du développement endogène est peut-être le plus difficile à combler, en particulier lorsque l'agence qui apporte les fonds est un acteur mondial. Nathalie Tinguery, coordinatrice du programme de pays pour l'US African Development Foundation (USADF) au Burkina Faso, réfléchit à son expérience de l'incorporation des valeurs et des objectifs dans ses pratiques de développement auprès des communautés et pour un bailleur de fonds international. Elle décrit comment elle est restée concentrée sur le développement endogène, ce que cela signifie sur le plan de sa pratique de développement et quels sont les aspects des politiques et des pratiques de l'USADF qui rendent cette sorte de développement endogène possible.

No obstante la muy amplia brecha existente entre las teorías y la práctica real de desarrollo, la brecha existente entre los conceptos vinculados a enfoques endógenos y la práctica de desarrollo endógeno puede ser la más difícil de reducir. Ello resulta particularmente notorio cuando la agencia de financiamiento participante es un actor a nivel mundial. A partir de haber incorporado valores y objetivos obtenidos durante su práctica de desarrollo, Nathalie Tinguery, coordinadora nacional de programas en Burkina Faso para la Fundación para el Desarrollo EE. UU.-África (USADF), medita en el presente artículo acerca de la experiencia surgida en su trabajo con comunidades y con una financiadora internacional. La autora muestra cómo sigue centrada en el desarrollo endógeno, explicando lo que ello significa para su práctica de desarrollo y lo que al mismo tiempo revela de las políticas y de las prácticas de USADF al posibilitar esta modalidad de desarrollo endógeno.

Conceptual clarity for the practitioner

From the start, one needs to be clear about the concept of development that drives practice, and to return to this definition regularly. As a development practitioner, currently working for USADF[1] in Burkina Faso, my definition of endogenous development is simply community-led and controlled development. For my understanding, the development idea, the derived initiative and actions should be locally controlled while at the same time there is openness to integrating innovations. There are many definitions of endogenous development but I find Compas's definition particularly useful.[2] Compas sees endogenous development:

> "as a process of change that enhances local control of the development process and primarily builds on local resources. It makes a distinction between social, physical, economic, human, spiritual and produced resources. This approach thus explicitly recognizes the importance of cultural diversity and suggests that there are at least as many notions of 'development' as cultures. Endogenous development is understood as based mainly, but not exclusively, on locally available resources. It has the openness to consider, modify and integrate traditional and outside knowledge. It has mechanisms for local learning and experimenting, building local economies and retention of benefits in the local area. The main actors in endogenous development are the people in the communities, villages and towns, with their own self-determined traditional organizations and leadership and civil organizations that have emerged more recently. They are the main carriers of the development process."
> (Compas 2011, 12)

Why endogenous development in Burkina Faso?

Endogenous development is important for Africa, including Burkina Faso, as it implies a development grounded from the bottom up, a development which stems from within and which respects the choices of the promoter and beneficiary local communities. As this implies building on existing local knowledge, institutions, strategies, resources, and involving the real stakeholders, endogenous development seems to me a sound approach considering ultimately that people constitute the main driving force to model and achieve sustainable development.

The basis for sustainability lies in how well the development initiative is grounded in and appropriated by the beneficiary community, and in their engagement in the process of achieving the social, economic, environmental, and ethical dimensions of the desired change. As I ask myself why we should foster endogenous development in Burkina Faso, I specifically think about the efforts in the country to develop locally based and viable economic institutions in the agricultural sector. In a landlocked country, with scarce resources and with a majority of smallholders in the agricultural sector, who work small family plots of 0.5 to 3ha, endogenous development becomes important when it creates opportunities to improve income, create jobs, and remove constraints to productivity, local processing, and access to markets. Smallholders are key stakeholders and in fact the engine of economic growth and human well-being (see Europafrique 2013). It is imperative to take into account their knowledge, existing rules of traditional management of communal resources, and the place of the youth, women, and other specific groups in the farming of different crops in order to identify constraints and concerted responses. It is equally important to build their capacities and set up local multi-stakeholder platforms that attribute a role to various categories of people and allows their participation in decision-making. It finally takes the integration of innovations systems, which focus on access to appropriate technology, and the quality of service delivery of different stakeholders in specific value chains, as well as the synergistic actions.

Can you implement endogenous development when working with an international funder?

The approach of USADF is consistent with endogenous development as its process of project identification, development, monitoring, and evaluation is built around the pillars of communities' participation and empowerment. First, there is the principle to reach the most underserved communities, and to invest in their ideas and initiatives in order to improve the local economy and create jobs. USADF is also building the capacity of local consultants and technical support systems by insisting on African-led country programme management (USADF 2013). In the Burkina context, the most marginalised people are identified as women, the rural youth, farmers/or artisans in the most remote rural areas, and people with disabilities. In my role as Country Program Coordinator, I spread the word to those target groups who might not access information and/or who live in the most remote areas across the country. To minimise the barrier of illiteracy (which concerns nearly 30% of the adult population in 2007), an effort is made to orally explain the programme in local languages. These communities already have or are organised into farmers' groups, umbrella unions, cooperatives, and associations, which may apply to the programme based on their own interests and ideas for economic projects. What is expected from the organisation is to effectively consult its membership base through some form of meetings, such as a general assembly, to ensure that the project concept they submit for USADF funding reflects the interests of the majority. The USADF application template used to present the project idea has been simplified to make it more accessible to the type of grass-roots organisations we are focusing on. Applicants are allowed to hand write their submission and it is my role to seek additional information when clarifications are needed. Usually the most difficult part is helping them identify what amongst a multitude of needs is truly the core problem, and one that is within the realms of possibility to resolve.

Applications are screened against criteria, including the marginalisation profile of the group and their track record of working together on a project, the potential of its economic project, and how it aligns with the country priorities for food security, rural enterprise development, market demand of the product, and management capacities of the group. Following pre-selection of an application, a field visit engages the group and its membership in discussion to assure a shared understanding of the economic project of the group: what is the problem being addressed; what are the causes and consequences; can group members identify solutions; what do they expect from the proposed project, and what will they contribute? (USADF 2009).

It is critical to ensure that any proposal being developed reflects the interests of the majority and not only the board or the most influential people. If it is difficult to engage the participation of women, one solution is to meet them separately before a joint meeting is organised. In general, various tool kits for participatory methods are helpful. If discussions demonstrate:

- group cohesion and energy
- organisation registration
- a governance system that abides by its internal regulations
- willingness for all people to participate actively in decision-making
- basic management capacities
- potential profitability of proposed activity (e.g., domestic or export market demand)
- environmental sustainability
- equitable access to benefits

then the Burkina Faso office recommends the proposal to USADF headquarters. Though it may appear that underserved groups cannot have these qualities or meet such criteria, USADF working principles assume differently. Group members may be illiterate and live in villages, far from departments, provincial, and regional centres, but they still have the collective capacity to run a locally grounded initiative with their limited resources, skills, and knowledge.

When headquarters concurs that the group has the potential to make good use of USADF support, a local partner organisation (usually an NGO or consulting firm) team returns to the group, and using a participatory approach prepares the full package of the project with all the pieces including the estimated budget and work plan, financial assessment, market analysis, technology analysis, and complete environmental screening report (USADF 2009). During this period, there are back and forth exchanges between the applicant, the technical assistance team, and USADF. This is essentially a free service and learning for the group where they crystallise their goals and transform "what" they want to do into "how" they are going to do it.

Following the development phase, a grant is put in place and the contractual documents explained in the local language before being signed. While the grantee has the responsibility of project implementation including recruitment, procurement, and grant financial management, the multidisciplinary technical assistance team from the USADF partner organisation works with the grantee to ensure the successful implementation of the project. Monitoring the grant is a joint effort of the grantee, the technical team, and USADF, using field visits and subsequent meetings, emails, phone communications, and data collection to inform different project performance indicators. The frequency of monitoring depends on the capacities of the group, ranging from visits every quarter to ensure a good start up to every six months thereafter (USADF 2009).

Reviews of the country programmes are organised every six months and this is the formal occasion at the level of USADF headquarters to assess the progress of different groups and country portfolios in terms of achieving required performance on funds utilisation, outputs, and impacts. Sources of information are quarterly reports produced by the grantees and the performance assessment provided by the local team.

Applying endogenous development when dealing with an international organisation could be challenging unless there is a shared vision or approach to economic development. The international organisation has its strategic priorities, its own stringent procedures to allocate funding, and its reporting obligations to its contributors. Some donors fund projects based on consultation and negotiation with governments, which theoretically represent the best interests of the local communities. Others like USADF go directly to people while remaining aligned with the national development priorities. Whatever the design of the donor's intervention, it is important to stay attuned to the context of the local communities.

In my view, USADF is among those donors who align its interventions with the principles of endogenous development, specifically on the dimensions of what Niels Röling (2007, 102–103) has called *"developing stakeholdership, economic institutions and also integrating the innovations systems"*.

The process of planning and implementing USADF-funded projects helps to bridge the gap between the intent and reality of endogenous development in several ways. First, the steps of establishing a clear diagnosis of the situation are done *with* the applicant group and development of the project uses participatory approaches. This sets the stage for capacity building and decision-making as well as groups developing their self-confidence and "voice" to speak out about matters that impact their lives. Implementation, monitoring, and remediation are the responsibilities of the grantee, which can use *local* technical partner assistance. Solutions to problems come first from the grantee; when expert advice is sought, it should be understood and endorsed by participants.

Capacity building is key to the USADF approach. Most projects include organisational, technical, financial management, and marketing capacity building, delivered through locally available experiential training and technical assistance. Although a typical project is about production, processing, collecting, storing, and marketing, training often includes literacy if the group asks for this to improve their business management.

An example of USADF's bottom-up strategy, coupled with participatory problem-solving, leading to endogenous development can be seen in the Tin Ba Sesame Production and Marketing Project. Association Tin Ba pour le Bien-être et le Développement Rural ('Tin Ba') is an association established in 1999 in the Fada N'Gourma area, east region of Burkina Faso with the objectives of (1) reducing poverty in rural areas; (2) promoting health and education for all in rural areas; (3) assisting and protecting women and children in their rights; and (4) protecting the environment, particularly the conservation of biodiversity. After experimenting with different livelihood options, Tinba decided in 2008 to focus on improving the incomes of its estimated 1,700 members through the production and marketing of sesame. Sesame was viewed by the membership as an alternative to cotton, inputs for which are too expensive for many poor small farmers. Additionally, payment to farmers for cotton sold to marketing boards was frequently tardy. Traditionally farmers produced sesame for domestic consumption. Recently government policy has encouraged the production of sesame for export markets. Tin Ba experimented for two years with sesame production and marketing, demonstrating its financial profitability. The association then made a strategic decision to develop Tin Ba as a credible intermediary between the sesame farmers and the international market and drive the expansion of sesame production.

Tin Ba's approach to USADF for financial support was demand-driven. Tin Ba wanted to create the conditions that would stimulate farmers to produce higher quality sesame. To do this they needed to overcome constraints at organisational and individual farmer levels. These included: relatively poor organisational capacity and structure which impeded operations of different decision-making bodies; insufficient administrative and financial management skills; lack of technical knowledge about production techniques that would producer higher yields; lack of working capital to purchase sesame from members; and limited access to improved seeds and equipment.

Two years were devoted to building the capacities of farmer members and of Tin Ba's leaders and staff. Results of this investment are related to the following dimensions of endogenous development:

Developing stakeholders:

Tin Ba restructured itself through a participatory process with the farmers' members groups. Thus, 139 producer groups were aggregated into nine communal unions. Decision-making bodies were reviewed and adjusted to represent the different sub-groups. The different sub-groups got their leadership trained in administrative and financial management and cooperative management. Members developed increased awareness of their own roles and responsibilities and the rules of associative governance. Endogenous facilitators were identified and trained in community mobilisation and participation techniques and provided field monitoring services and quality assurance backup. Regular meetings of the decision-making bodies were used to anticipate implementation issues and to take remedial action. Tinba's assets can be identified as its strong leadership, the engagement of the farmers, and the commitment of all to move forward.

Developing locally based economic institutions:

To facilitate the process of change, USADF support included mobilising the expertise of a local consultancy firm to provide technical support to the group. Such technical assistance impacts both on Tin Ba as the umbrella organisation and on the farmers organised into groups to improve the quality of the product delivered to clients. The two-year capacity building phase put in place a working capital fund for the purchase of the sesame production and allowed farmers access to improved seeds and to testing of seeds in order to increase yields. Equally, a number of tools were set in place to strengthen Tin Ba's capacity to be a credible development actor and to negotiate with buyers and also mobilise additional funding through financing institutions. These are a manual of procedures, a business plan, the installation of management tools, and an accounting system to put in place conditions of transparency and accountability.

The capacity building project led Tin Ba to build a stronger relationship with members by delivering good quality services to them, including seed provision, fertilisers, training, purchase and marketing of sesame, and timely payments. Tin Ba progressively increased its marketing capacities and offered fair prices to the farmers while paying them in a timely manner. These actions engendered more farmer loyalty. Product revenue increased through higher yields recorded and improved quality of sesame. These visible results maintained farmers' interest in producing more quality sesame and to giving priority to Tin Ba above other potential purchasers. Today, Tin Ba is ready to implement its business plan negotiated with the different stakeholders and thus to become a major player in the sesame value chain. The achievements during the first project permitted them to get a follow-on Enterprise Expansion Investment grant from USADF. This second grant aims to build storage infrastructure, increase the working capital for seeds/fertiliser, set up the farmers' equipment fund, acquire sesame cleaning and transport equipment, and conquer in a more proactive way the export market. The emergence of the farmers organisations into stronger entities and the change of Tin Ba into a marketing cooperative have been identified by Tinba as the way forward.

Managing the tension between donor needs for immediate results and capacity building?

It is true that donors need to see some results over the short term that they can show to taxpayers who fund development projects. But those of us working in development know that miracles cannot be achieved after only two to three years of support. That is why we usually approach results on two levels. The first level is to help the group get the necessary organisational capacities and minimal management systems in place, as well as some experience of collective management of working capital, equipment, and infrastructure. In general, the largest part of initial funding goes on training opportunities for as many members as possible. These types of projects are typically for two years. For this capacity building, we track indicators for increased outputs like crop yields, revenues, and membership increase. This is because we are in the sector of enterprise development. We also look at more qualitative results related to management and governance improvement, enhanced credibility, and self-confidence of members.

If the group proves viable and its proposed enterprise offers good potential (serving a booming market) it may decide to seek further support from USADF for infrastructure, equipment, and working capital to expand the enterprise. The expansion project may be for three to five years, giving the time to see impact achieved over a longer period. Measuring impact may be against quantitative indicators such as increased yields and earnings, as noted above for Tin Ba. More important are measures of local group capacity to sustain the enterprise and expand it once the outside support is finished, for example, whether the group is able to negotiate directly with a commercial bank to meet future investment needs such as working capital and other capital expenditures.

Final thoughts

This note refers to capacity building. But more than that, endogenous development is about *capabilities* building. We strive for independent groups that can free themselves from grants and run their business in a self-sufficient way where their operational costs are covered and they could secure benefits. The micro, small and medium size enterprises we help develop generate benefits for individual members in the group, create jobs for their own communities, and enable community-led and controlled development.

Notes

1. The United States African Development Foundation (USADF) is an independent federal agency established to support African-designed and African-driven solutions that address grassroots economic and social problems. USADF's unique mission and structure makes it possible to carry out development in isolated and highly marginalised regions in Africa.
2. COMPAS (COMPAring and Supporting endogenous development) is a capacity building programme to develop and mainstream endogenous development methodologies. Field programmes serve to develop evidence of the outcomes and impact. Mainstreaming takes place by including endogenous development in programmes funded through various donor agencies, establishing policy dialogues, and developing university curricula.

References

Compas. 2011. "Learning Endogenous Development." Leusden: Compas. Accessed February 5, 2014. http://www.compasnet.org/blog/wp-content/uploads/2011/03/Lendev/001-VanVe.pdf%20

Europafrique. 2013. "Family Farmers for Sustainable Food Systems. A synthesis of reports by African farmers' regional networks on models of food production, consumption and markets." Rome: Europafrique. Accessed August 3, 2013. http://www.europafrica.info/fr/publications/les-agriculteurs-familiaux-luttent-pour-des-systemes-alimentaires-durables

Röling, Niels. 2007. "Endogenous Development and Resilience: The Institutional Dimension." In *Endogenous Development and Bio-Cultural Diversity: The Interplay of Worldviews, Globalization and Diversity*, edited by Bertus Haverkort and Stephan Rist, 101–115. Leusden: Compass.

USADF. n.d. "Policy Manual MS 232 Project Quality Assurance." Property of USADF. All rights reserved.

USADF. n.d. "Program Guidelines." Property of USADF. All rights reserved.

USADF. 2009 "Training Manual for Country Program Coordinators": "Project Selection, Review, and Approval Guide". Property of USADF. All rights reserved.

USADF. 2013. "Tin Ba application to USADF and the CPC site visit report, 2013." Property of USADF. All rights reserved.

USADF. 2014. "USADF programs." Accessed June 30, 2014. http://www.usadf.gov/programs.html

Donors and exogenous versus endogenous development

Susan H. Holcombe

Economic and social development has occurred through the millennia. The post-World War II and post-colonial periods have ushered in a new era of donor-led development assistance policies and of professional development practice. Since the 1950s, development has been conceived of as rich and technologically advanced countries helping poor countries develop – a delivery system of development. The dominant development priority has been economic growth as opposed to livelihoods and social/human development. With some interesting exceptions, development, seen as development assistance, has been largely top down, or *exogenously* driven. In recent decades, scholars, practitioners, and even donors have called for participatory approaches, building on local institutions, culture, capacity, and local ownership. There remains a gap between the rhetoric and the practice of development. If we want to move from top-down, exogenous development to development that encompasses *endogenous* approaches, we must understand the barriers posed by practices in both donor and recipient country development organisations. This paper explores the barriers and possible remedies open to practitioners, policymakers, and academics at different levels.

Le développement économique et social a eu lieu au fil de millénaires. Les périodes post-seconde Guerre mondiale et postcoloniale ont marqué le début d'une ère de politiques d'aide au développement impulsées par des bailleurs de fonds et de pratiques de développement professionnelles. Depuis les années 1950, le développement a été conçu comme une situation dans laquelle des pays riches et technologiquement avancés aident les pays pauvres à se développer – un système dans lequel le développement est « fourni ». La priorité première en matière de développement a été la croissance économique, et non le développement de moyens de subsistance et le développement social/humain. Sauf quelques exceptions intéressantes, le développement, vu comme l'aide au développement, a été largement directif, ou impulsé de manière exogène. Au cours des quelques dernières décennies, les théoriciens, praticiens, et même les bailleurs de fonds ont promu des approches participatives, basées sur des institutions, la culture et les capacités locales, et sur l'appropriation locale. Il continue d'y avoir un écart entre la rhétorique et la pratique du développement. Si nous voulons nous éloigner du développement exogène directif pour aller vers un développement qui englobe des approches endogènes, nous devons comprendre les barrières que font surgir les pratiques dans les organisations de développement des pays donateurs ainsi que des pays récipiendaires. Cet article examine les barrières et les remèdes possibles disponibles aux praticiens, aux décideurs, et aux universitaires à différents niveaux.

El desarrollo socioeconómico ha sido impulsado a lo largo de varios milenios. Una vez terminadas la Segunda Guerra Mundial y la era colonial, se abrió un nuevo periodo en el cual los donantes empezaron a marcar la pauta de las políticas de asistencia para el desarrollo, surgiendo, paralelamente, una práctica de desarrollo profesional. Desde los años cincuenta, el desarrollo ha sido concebido como una ayuda brindada por los países ricos y

136

tecnológicamente avanzados a los países pobres, es decir, como un sistema que proporciona desarrollo. El mismo ha dado prioridad principal al crecimiento económico en lugar de al desarrollo social y de los medios de vida. Salvo contadas e interesantes excepciones, se ha dado al desarrollo un carácter asistencial, en gran parte vertical, impulsándolo de manera exógena. Sin embargo, en décadas recientes, los académicos, los operadores e incluso los donantes han hecho llamados con el fin de que se promuevan enfoques participativos, construidos a partir de instituciones, culturas y capacidades locales, que impliquen la apropiación local de las actividades. A pesar de ello, continúa existiendo una brecha entre la retórica y las prácticas de desarrollo. Si el objetivo es caminar desde el desarrollo exógeno y vertical hacia un desarrollo que incorpore los enfoques endógenos, es necesario comprender las barreras que representan las prácticas llevadas a cabo por las organizaciones de desarrollo, tanto aquellas de los países donantes como las de los países receptores. El presente artículo examina dichas barreras, señalando las posibles soluciones que se presentan para los operadores, los formuladores de políticas y los académicos en los diferentes niveles.

Introduction

Most voices in this special issue are those of scholars and practitioners from Africa. They discuss endogenous development from two perspectives: (1) participation, local capacity and knowledge, and ownership; and (2) a deeper level of using the historical and cultural experience of a people to inform the goals and shape the path of development for that society. One can argue that endogenous development is not a new concept or set of values about what development is and how it is done; development over the millennia has had local ownership (if not full participation) and has evolved within historical parameters and local contexts. When we ask about donors and endogenous development we introduce the particular relationship between the giver and the receiver. While it is true that some countries (China) have developed without significant outside aid, and recognising the current critiques about the efficacy of aid (Easterly 2002; Moyo 2009), we still make an argument that a transfer of resources from the rich to the poor has the *potential* to make a difference in the lives of people in developing countries. Therefore we do need to ask how and whether the exogenous model works.

Inherent in development assistance is a power imbalance. Endogenous development implies a power shift from the donor world, which has controlled the terms of the development dialogue (see Escobar 2004), to societies seeking to control their own development. There are a range of voices in the donor world calling for variations on endogenous approaches to development. European scholars and practitioners, more than those in North America, have developed a body of research on endogenous development, sometimes with utilitarian objectives (sustainability) and sometimes in terms of identity, confidence, and dignity (see the work of Niels Röling; and the website of Compas: www.compasnet.org). Donor development agencies have incorporated participatory approaches and expressed intentions to build capacity, ownership, and respect for the local context into organisational policies and sometimes into ways of working. At the state donor level, agencies have advocated that development, to be sustainable, needs to reflect local priorities, practices, and culture, and to be locally owned. The Development Assistance Committee (DAC) of the OECD is working with recipient countries to articulate common commitments to implementing principles that reflect some aspects of an endogenous development approach. The Paris Declaration (2005), for example, commits donor and recipient country support for endogenous leadership (ownership) at the state level: it says that partner countries should "*exercise effective leadership over their development policies, and strategies and co-ordinate development actions*"; and development cooperation should be built "*on partner countries' national development strategies, institutions and procedures"*, if not precisely on locally rooted culture, context, and knowledge (OECD 2005, 3).

The challenge to putting these seemingly overlapping ideals into practice, this paper argues, comes from the power imbalance implicit in donor-recipient relations; the history of development assistance, the systems and procedures that continue to influence donor recipient relations; and the difficulty state actors have in engaging with endogenous approaches. There are ongoing frictions over certain questions, such as whether human rights are universal in view of local values that seem to conflict with international norms. Equally important are barriers such as recipient country leadership, corruption, and capacity. These tensions and power imbalances can be seen in provisions of the Paris Declaration, particularly the conditions on ownership and reliance on a country's strategies, institutions, and procedures, and on the call for results. For example, the Paris Declaration describes the decision to rely on a country's institutions and systems as dependent on whether *"these [institutions and systems] provide assurance that aid will be used for agreed purposes"* (4). There is probably little theoretical disagreement among DAC donors and most recipients about the need for accountability and for results. Disagreements flow from questions about whose accountability, who controls the indicators, and for what period of time. Finally, barriers derive from our varying definitions of "endogenous" and "development".

The voices in this volume are primarily the voices of Africans seeking to visualise how African history, culture, and leadership can shape development from an African perspective. This article is the voice of one Western development practitioner, now turned academic, reflecting on the largely post-colonial development experience of the past 60 years; asking what links to endogenous development exist among evolving donor theories and practice of development, and where the opportunities lie for supporting endogenously conceived and led development in Africa.

This paper explores "endogenous" and "development" in the context of the West's own development; the evolution of Western attitudes about how development takes place; and how barriers to endogenous development operate. Then it suggests how donors and recipients, in certain contexts, can adapt or strengthen capacities as well as management, incentive, and measurement systems, to foster endogenous development. Finally, it acknowledges forces not addressed in this paper, which need separate attention.

Definitions

We, as Western development practitioners, may start with an agreement that endogenously-defined development is a process that represents a set of values, and it is also utilitarian in that it can be a demonstrated benefit in terms of sustainability of development interventions. This would seem to be the intent of some provisions of the Paris Declaration. However, this definition needs to go further.

"Endogenous" is an off-putting, loaded, and overly complicated term. In biology, endogenous refers to substances or processes originating from within. It is Greek in origin, meaning "growing or developing from withing" (http://dictionary.reference.com/browse/endogenous), and is neatly distinguished from exogenous, which means coming from without. To expand the biological definition further, exogenous factors may be both beneficial (therapies) and harmful (viruses); likewise endogenous factors may be beneficial (circadian rhythms that regulate) or harmful (cancers). To complicate matters, endogenous and exogenous factors may influence each other and mutate. Unfortunately (or fortunately), social behaviours are more complex, and though these terms may clarify in biology, they too often confuse in social sciences.

As this volume suggests, there is a range of definitions of endogenous development and what this means for development policy and practice. For this essay the definition of endogenous development has three facets. First, endogenous development represents locally-defined,

-led, and -controlled efforts to expand human choice. This encompasses the notion of owner-ship, but also puts a value framework on development that is influenced by Amartya Sen's freedom, capabilities, and agency approach. We can remember that for Sen (1999, 3), devel-opment is about *"expanding the real freedoms that people enjoy"*, or as the UNDP (1990, 10) phrased it in its first Human Development Report, development is *"a process of enlarging people's choices"*. Sen posits five freedoms – political, economic, social, transparency, and protective security – which are both ends in themselves, but also the means to expanding free-doms. The second facet is also related to Sen, and this is the notion that development is about human dignity and self-respect. Sen talks about things or commodities as being a means to an end, and that having certain possessions or types of clothing are necessary to being able to participate, or, quoting Adam Smith, having *"the ability to appear in public without shame"* (Sen 1999, 72). If development assistance, implicitly or explicitly, is an aid delivery system, assuming an inequality between the giver with superior assets and technology, and a receiver with capacity gaps to be filled, then the recipient country stands, to expand on Sen's metaphor, as a person in tattered and dirty clothing in a gathering of suited and bejewelled people. Mohammad Yunus, founder of Grameen Bank, has given the example of how poor people, and particularly poor women, do not feel comfortable entering a formal bank and find banking procedures alien (Holcombe 1995). These examples suggest that organisations and institutions controlled by the forms and processes of those with power can exclude those without power. The very nature of donor procedures may exclude. The exclusion of the powerless is more than physical; it is psychological and denies dignity to the excluded. This observation leads to the third facet of endogenous development, which is related to the first: in being locally-defined, -led, and -controlled efforts to expand human choice, endogenous development is rooted within the particular context and culture it serves. Adap-tation and learning from outside in order to expand human choice are possible and desirable, but change is led from within, not imposed from without. This places a heavy responsibility on leaders of endogenous development approaches, but from the exercise of the responsibility building on (but not exploiting for political purposes) one's culture and traditions comes the dignity referred to above.

Rajesh Sampath, who is exploring *"the philosophical and religious underpinnings of devel-opment"* (personal email communication, July 17, 2013), looks at the similarities and contrasts between the dominant paradigms of the West, versus those of other cultural regions. One of the reasons that thinking about endogenous development is so difficult is, as Sampath suggests, that

"Philosophical and religious roots run so deep that they are barely perceptible when working within the epistemological frameworks of the social sciences and social policy." (Sampath, personal email communication, July 17, 2013)

Thinking about endogenous development in Africa is even more difficult, because the philosophi-cal and religious traditions are diverse, largely oral and, when written, often written by outsiders who risk bringing their own philosophical biases.

Barriers to adopting endogenous approaches
Faulty memories of Western development
It has been easy for the West to glamorise and to simplify the history of its own development as a basis for believing that its models of democracy and capitalism are best. This is scarcely the place to review the history of Western industrial, social, and political development, but a few

observations may provide some useful context. Western development took place in a privileged context. It had the time, space, and resources to develop over a long period and according to its own, endogenous patterns. The technology for communication was limited, giving space for local development. Migration was not controlled. Globalisation as we know it today did not exist. Natural resources were locally exploited. The emergence of political participation and accountability emerged slowly; the Magna Carta in 1215, in which barons wrested powers from the monarch, is one example of the slow accretion of the rights to participate. Until recent centuries, GDP per capita was higher on the Indian sub-continent than in Britain. Europe languished in the middle ages while science and philosophy flourished in the Arab world. As Europe industrialised in the nineteenth century, emigration to the New World was an option for unemployed or dissatisfied labour. The US, during its period of rapid growth, violated most of the rules of the Washington Consensus[1] which, through the IMF and World Bank, the US has sought in recent decades to impose on developing countries. James Weaver, Professor Emeritus of Economics at American University, notes how the US violated the rules during its period of growth:

> "The U.S. developed behind high tariff walls to protect its infant industries throughout the 19th and much of the 20th centuries. Secondly, the U.S. stole the technology to start its industrialization from Europe as immigrants brought the ideas and plans for textile industries ... Third our financial system was based on unchartered, unregulated and unsupervised banks. Our economy was made great by unsound finance. Fourth, our governments were riddled with corruption, at all levels ... " (Weaver 2002, 1)

A glossy version of the history of Western development may allow us as Northerners to forget that it took time to develop the economic and political governance and underlying institutions that support modern economies. These romanticised memories may obscure the challenges of poverty, inequality, and exclusion that remain serious in many OECD countries, despite aggregate high levels of GDP per capita.

Western aid and development philosophies in practice: top down not bottom up? Mechanical versus organic?

As we discussed endogenous development, the co-editor to this volume, Chiku Malunga, suggested that:

> "the donor system, as it operates now, seems to be a machine – something mechanical. The arguments for endogenous development seem to be [searching] for a natural metaphor, like a tree or a plant or some animal ... the question I [have] is what is it going to take to change the metaphor from a machine to a living system? To whom are we talking and to whom are they listening?" (Malunga, personal communication, August 27, 2013)

This coincides with the observations of a recent publication that comes from Northerners who have worked with communities for many years. The Listening Project sought the opinions of more than 6,000 people who had received international assistance. The resulting online book, *Time to Listen* (Anderson, Brown, and Jean 2012), notes the surprising consistency of people's descriptions of their interactions with the international aid system. They spoke of aid becoming a business, focused on delivering an input thought to produce change; they characterised aid as being top down, focused on a project cycle (Anderson, Brown, and Jean 2012). They agreed on the over-bureaucratisation of aid that diverts energies from impacts on people to complying with forms and reporting requirements:

"We have coined this quite awful word 'proceduralization' ... to describe the codification of approaches that are meant to accomplish positive outcomes into mechanical checklists and templates that not only fail to achieve their intent but actually lead to worse outcomes." (Anderson, Brown, and Jean 2012, 66)

Dominant Western philosophies and approaches to development, it can be argued, are top down, monocausal, and primarily focused on economic growth – which is relatively easily measured. The consequence of this has been a series of development theories that determine the approaches of major aid donors, and lead to a focus on achieving results of the intervention without necessarily considering the interests of recipients.

The economists' domination of development and donor agencies has been strong and continuing, even when the utility of economic theories has been called into question. It led to the search for single causes of underdevelopment and thus the one critical intervention that would produce growth – the X that will produce Y. This monocausal approach, as Irma Adelman (Adelman 2001) called it, has characterised the different approaches that dominated donor agency practice since the 1950s. One can think of the input output theories of the 1950s, where missing investment would spur production and growth, or of the theory that technology would provide the missing ingredient that spur growth. More recently, the Washington Consensus promised economic growth, if only the country in question would follow a set of policy prescriptions designed to restructure and open the economy to market approaches and trade. Each of these policy interventions was going to deliver growth, and each of them was so deeply embedded into donor agency practice that abandonment of any one approach is difficult. Easterly (2002) noted that input-output, aid to investment theories of growth, though disowned even by the economists (Harrod-Domar) who articulated them, were still part of World Bank appraisal procedures. Easterly[2] and Rodrik have noted the failure of Washington Consensus policies to produce promised economic growth, with Rodrik (2002, 2006) observing that countries that did achieve substantial growth in the 1990s did so by violating some of the prescriptions of the Washington Consensus while developing their own locally-derived solutions to meeting the requirements of growth.[3] Though some have asked whether the Washington Consensus policies and the prioritisation of economic growth are dead, in fact the issue remains alive in the discourse among development thinkers. In August 2013 economist Jagdish Bhagwati challenged Amartya Sen's assertion that India has failed to address poverty and inequality because of failure to invest in human development. The *New York Times* reporter Gardner Harris (2013) wrote that, to Mr Bhagwati, *"India's myriad problems have less to do with poor health and literacy than a poor investment climate. Give people more jobs and money and they will invest in their own education and health"*. It can be argued that even a heartfelt focus on ending poverty can result in the essentially top down approach of the Millennium Villages Project implemented by Jeffrey Sachs and his colleagues.

The donor world may never have critically examined whether these top down, monocausal approaches could actually be implemented or achieve intended purposes. There is little incentive, and considerable disincentive, for the donor world to acknowledge failures. NGOs, bilateral aid agencies, international organisations all need to show that aid has been successful so that they can convince the funders – parliaments, taxpayers, and individual donors – that more money should be invested. In reality, much donor money has continued to flow, particularly to national levels, despite evidence of failure to achieve results. Easterly discussed the flaws in the input-output theory of growth and the development approach of filling the financing gap (input) to achieve economic growth, both of which guided the thinking of international financial institutions. He gave the vivid example of Zambia:

"a comparison of what Zambian's actual average income would have been, $2 billion of aid later, if filling the financing gap had worked as predicted ... Zambia today would be an industrialized country with a per capita income of $20,000, instead of its actual condition as one of the poorest countries in the world with a per capita income of $600 ... " (Easterly 2002, 42)

Easterly's point is that donors continued the investment in Zambia over more than 30 years, despite the absence of evidence that such investment in financing made a difference in per capita income. A public administration scholar might term this *path dependence*. Breaking out of that path is difficult. Admitting mistakes may threaten the viability of the aid industry. Political considerations, not effectiveness, may determine the directions of aid.

Despite the growing critiques of aid effectiveness over the past decades (Easterly 2002; Ayittey 2009; Moyo 2009), there has not been a fundamental rethinking of aid among donor organisations. A most significant reaction has been an increasing emphasis on managing and measuring the results of aid. In principle there can be little disagreement with the importance of measuring results. It should be not simply an obligation to donors, but a responsibility to the people in whose name the aid is received. In practice, the emphasis on results may lead to mechanical attention to reporting outputs and accounting for inputs, and actually degrade the strengthening of endogenous capacity to manage aid.

Both donor and recipient countries, in the Paris Declaration, called for *"managing and implementing aid in a way that focuses on the desired results and uses information to improve decision-making"*. Recipient countries agreed to " *endeavour to establish results-oriented reporting and assessment frameworks that monitor progress against key dimensions of the national and sector development strategies; and that these frameworks should track a manageable number of indicators for which data are cost-effectively available"* (OECD 2005, 8). Donor nations also agreed to support capacity building. The challenge of results-based management comes with the implementation (OECD-DAC 2005). Most aid is project-driven and the resources for measuring exist within the timeframe of the project only. Donors are under pressure to report results to their funders – parliaments and taxpayers. An NGO leader in Lebanon reported to the Listening Project:

"For international donors, a project is only useful if it has immediate results that they can show and measure. How can you heal a trauma in six months?" (Anderson, Brown, and Jean 2012, 80)

Mohammad Yunus has suggested that the real results of small-scale savings and lending for women may be seen, not in short-term changes in assets and income, but in intergenerational changes in skills, knowledge, and a new belief of people coming from poverty that they can make a change in their lives and communities (Holcombe 1995).

Reporting results to donors and funders is important for receiving the next tranche of aid. Observers of the process note that frameworks for monitoring results can become complicated, divert staff time and energies, encourage fraudulent reporting, and may actually miss capturing the important changes. The complexity of monitoring for results may limit the kind and degree of participation of local organisations and communities (Anderson, Brown, and Jean 2012). The capacity building and institutional changes that underpin sustainable development may require decades to develop – as they did in Western countries during their periods of industrial development. Fukuyama (2004, 30–40) suggests that these capacities and institutional and cultural changes are little influenced by external aid and conditions, but they result from demand inside the society in questions. A focus on results, within the time constraints of the project cycle, to the extent that it leads to a heavy "proceduralization" of development assistance, drives out new thinking about how to build the demand in developing countries for better institutions or a competent bureaucracy.

African environment for endogenous development

Voices from Africa may call for endogenous approaches to development, meaning variously African-defined and led development or more specifically development, particularly at the community level, that is based on community values and customs. There are multiple barriers within the African environment to moving towards any of these dimensions of endogenous development. Key among these are weak governance at all levels, and the focus of endogenous development on the local as opposed to the national.

Governance in Africa is an enormous topic, but it is fair to say that historical, political, natural resource endowment, and globalisation factors have contributed to poor leadership, corruption, and weak ability to implement the stated intentions of government. This has negatively affected the capacity of governments to control their own development trajectories and manage their relationships with outside donors. Upon independence African governments were weak and focused on survival, not development. Claude Ake noted that "*the struggle for power was so absorbing that everything else, including development, was marginalized*" (in van de Walle 2001, 116). Van den Walle (2001, 116) argues that post-colonial governments, in these circumstances, had a "*style of rule that ... combined authoritarian legacy of colonial administration and the village traditions of patrimonialism*". Patrimonialism sometimes led to a form of clientelism where leadership rewarded their own ethnic group and other supporters (at the expense of others) in order to preserve power while seeming to honour the requirement of democratic elections. In cases where leadership adopted African symbols, the adoption was intended to legitimise authoritarianism, not build on traditions of governance and social relations (van de Walle 2001, 113–137).

If one surveys African governance today there is evidence that governance is improving, the rules of government are adhered to, and that increasingly citizens may attempt to hold leadership accountable; but progress is slow, particularly when seen in the context of the demand for results. Institutional development was seen by donors as one way to rectify the failures of the Washington Consensus. Institutions reflect the informal and formal ways in which decisions are made and action taken in the political or economic spheres. This kind of change takes place from the inside out, and as a result of the demand of people to change institutions; and it happens over a long period of time. Donor conditionality on governance reforms may introduce new structures, for example under decentralisation reforms that are intended to increase decision-making at the local level. Unless the reforms are accompanied by changes in the formal and informal institutions that support decentralisation, this kind of decentralisation is unlikely to result in improved governance.

Globalisation has many influences on the African environment for endogenous development. The obvious threats to endogenous approaches come from commercialisation and the emergence of a global, material culture. Behind the power of commercialisation are the major corporations that control not just financial and product flows, but also, by virtue of their size and power, considerable influence over governance. Michael Chege (2008), writing about governance in Kenya, noted the high incidence of corruption and the impunity of corrupt officials, identified as one example the Anglo-Leasing Scandal in 2005. John Githongo, the Kenyan official responsible for investigating and punishing corruption, documented the involvement of senior members of government in this scandal. As a result of death threats he was then obliged to flee the country. Similar stories are repeated elsewhere. The point is that there is no effective global governance of multinational corporations, meaning that corporations can engage in bribery of African governments with negligible fear of suffering penalties themselves. This plays out in the extraction of natural resources, rents from which have led to corruption of officials and poisoning of governance processes in most resource rich countries in Africa (see for example Collier 2007; Diamond and Mosbacher 2013).

Governance is weak in most African countries, so that there is little leadership available to wrest control of the development agenda from the donor community, or even to consider how to build on legitimate history and culture to forge new states and strong institutions. The power of multinational corporations, particularly extractive industries, contributes to the continuing corruption of Africa governments.

Bridging barriers

It is far easier to say what is wrong with donor approaches than to suggest remedies. In part this is so because some of the remedies (global governance of extractive industries or ending Northern aid to authoritarian leaders in Africa that has been maintained for economic or political purposes), are not going to happen directly. In part, remedies need to be grounded in endogenous leadership and based on a sense of confidence in African dignity, integrity, and ability to lead change in Africa. This article makes recommendations for donor agencies for investments that have the potential to lay the foundation for longer term changes in how Northern aid is designed and implemented. These modest steps are intended to create incentives for a different type of aid, and a mindset and capacity in both North and South to create the conditions that encourage endogenous development and endogenous leadership of development.

What we can do as Northerners: to rethink our own mindsets

An example can help illustrate what this means. A call for papers to be presented at a conference on aspects of *global* development was circulated in 2013; it listed 135 of the confirmed speakers for this conference. Of the 135 names, only seven appeared to be names that might come from Africa, Asia, or Latin America, and several of these were based at institutions in the North. If we say we want participation and ownership from recipient countries then we need to do more than issue invitations. We need to go beyond invitations and make sure that these voices are *enabled* to come to the table, and that the table is not designed by Northerners alone. To the extent that the development dialogue is carried on at global conferences or in development journals, voices from the South may not appear because of lack of funds for travel; lack of access to basic literature and journals; or because of unfamiliarity with Northern norms that govern conferences, research, or journal submissions. Civil society organisations in the South, seeking to engage with development policy level at the national or international level, often lack the human and financial resources to develop research findings that can be persuasive against the research and presentations of global and bilateral institutions. The norms of development work are set in the North, and this is a reflection of the power imbalance. Efforts like the Listening Project, mentioned above, or the 2000 *Voices of the Poor* (World Bank) have sought to redress the power imbalance and create a space at the table. More broadly, changes in mindset are needed amongst development practitioners and organisations.

Invest in education and deployment of development practitioners and policymakers

Human capital in Africa today is much deeper and broader that it was in the 1950s, when much of the population was illiterate in any language and there were few university graduates. There is a growing body of Africans, educated in Africa and/or abroad, who have the skills, commitment, and cosmopolitan perspective, along with a rootedness in their own local and national communities, to provide endogenous leadership in government, the private sector, and in civil society.

We see today many examples of a new generation of Africans making change, but also creating an environment for change. Patrick Awuah started a values-based private university aimed at building African business leadership through education in critical thinking and problem-solving skills combined with commitment to ethical standards. Some, like John Githongo of Kenya, have taken heroic and even dangerous efforts to identify and combat the corruption that so weakens governance in Africa. The University for Development Studies in Northern Ghana was created to provide African leadership at the everyday, practitioner level as well. A key part of their programme is a field trimester requirement that sends students to villages to engage with communities in a joint problem-identification and -solving process.

The role for Northern governments and donors is to continue existing and to expand investment in education for Africans; yes, through fellowships abroad, but also increasingly through greater support to African universities, particularly when these universities have a commitment to reform. The African university systems were largely formed under colonial direction and many suffered deterioration during periods of structural adjustment reforms and politicisation. Donor funding can be used as seed money for supporting reforms that introduce critical thinking and a problem-solving approach and strengthen teaching of the humanities, the sciences, management, entrepreneurship, and of African history, culture, and traditions. Donors can support faculty research in Africa that strengthens local capacity. For example, as one practical note in this issue describes, faculty at the UNAAB in Nigeria developed a simple technology for inexpensive drying of cassava waste to remove toxins, thus making it suitable for animal fodder that accelerated the development of livestock. This was a locally developed innovation that appears to have worked because the faculty also understood the social conditions that allowed the innovation to be adopted and sustained by growing numbers of cassava producers and herders. There is African capacity in African universities to address challenges of economic production. Instead of favouring technical innovations that come from abroad, donors can search for African innovations, investments in which may have the double benefit of being rooted in the local context and of strengthening African organisations and capacity. It also moves toward a bottom-up approach.

The education of Africans abroad has offered them the advantage of stepping outside their own societies and gaining a more cosmopolitan vision. Many fellowship programmes have sponsored Africans for study abroad. While some of these students have stayed abroad and are part of a growing African diaspora, a significant cohort has returned. Return comes with its own limitations. The new graduate may come back with few assets and connections, and establishing or re-establishing a career is difficult. Moreover, idealism and aspirations for justice, equity, and good governance may erode when confronted with governance and economic realities. Return can be lonely. Educating Africans is not sufficient. There is a role for continuing donor support to returning graduates, either in terms of forming graduate associations and other connections, or in terms of seed funding to launch enterprises. Some fellowship funders are attempting this, but it needs to be done more explicitly and broadly.

Equally important is the education of development practitioners in the North. Schools of development studies in the North need to re-examine curricula – not only the knowledge that is being taught but also the value assumptions behind the content. The legacy of top-down approaches and a conviction that we can find the missing variable that produces development still permeate education. Even when the stated values of educational programmes are built around participatory approaches and inclusion, practices may carry implicit biases of Northerners knowing best and of bringing not only financial resources, but also the ideas and technologies. In an interview with the TedBlog about Dambisa Moyo's book, Dead Aid, Ghanaian economist George Ayittey referred to the *"presumption that Africans don't know what is good for them and that Americans or other foreigners know what is best for Africans is extremely offensive"* (http://blog.ted.com/2009/04/09/ayittey_on_dead_aid/). This perhaps reflects how many Africans actually see Northern development practitioners, despite the good intentions these workers may bring. Education, therefore, is not just about concepts and skills, but also the critical self-appraisal that permits us to change or open our mindsets. Without detailing here what a development studies curriculum that encompassed endogenous development might look like, one can suggest that more attention needs to be given to encouraging students from the North (and the South) to evaluate critically the biases and vision they bring to development work and to encourage the interpersonal and management skills that allow them to put this self-knowledge into practice.

Change our approach to measuring "results"

A common saying with respect to measuring is that "what gets counted, counts". In development, when the counting gets done it may exclude the important variables about local ownership, capacity, and sustainability because these are hard to measure. The way we measure results affects organisational behaviour, as discussed below, but it also keeps the focus on the limited timeframe of the project. A major reform, particularly for larger donors, would be to increase the focus on longer-term tracking of change, and to examine the process of implementation in a qualitative manner. To give an example, one could look at efforts to introduce technical innovations with the potential for positive human impact. It is important to know whether three-monthly treatment of worms not only improves school performance, but is more cost effective than other interventions. If we are interested in sustainability, however, the more important questions may be about whether and how communities and parents begin using de-worming medicine, using their own funds or demanding that this be a part of social services. Administering de-worming medicine is something that is done to children. Sustainability of de-worming measures implies an increase in capacity of parents and communities to make decisions for their own family well-being. Technical evaluations of the results of de-worming on school performance are simple and rational. Behaviour change is not always a rational process. Figuring out how a particular community appropriates an exogenous technical advance is difficult, messy, and perhaps context-specific, but studying the process of implementation across contexts may provide some generalisable lessons. Looking at whether a donor support intervention has been sustained and perhaps expanded over a 10- to 15-year period may be a better measure of the accountability of aid effectiveness. Measuring results must balance the assessment of outputs (short-term results) on one hand; and outcomes and impacts (long-term results) on the other hand. From practical experience, outputs may be measured within a year and half; outcomes may be measured within three years. Impacts may rarely be measured within five years. Impact assessments that are conducted immediately after a five-year project has closed or is about to close may not give a realistic picture of the impact and its sustainability. Conducting an impact assessment five to 10 years after a project has closed would give a more realistic picture of the project impact and its sustainability beyond the life of donor funding. This type of approach, however, is alien to most donor agencies' procedures.

Recognise that the challenge of endogenous development is about the organisations, structures, and systems that implement aid and change

Donors pay lip service to local ownership and participation, but aligning practice to stated goals requires organisational changes. For example, explicit and implicit incentives offered to development workers largely determine where their efforts are applied. Procedures prescribe how things are done, what steps the development worker must take. When the logical framework approach to project planning was introduced, it was intended as a way of aligning resources with expected outputs, intended outcomes, and goals. It was never intended to be a straitjacket on adaptation and innovation. The Listening Project suggests that, in practice, these procedures have driven out both opportunity for creativity and the focus on goals. Individuals and organisations are held accountable, if at all, for deliverables in the context of the project. There is no examination 10 years later about whether the project was sustainable. There are few rewards for breaking the pattern, but there are some examples of organisations attempting this.

The International Fund for Agriculture Development (IFAD) is one agency that has given thought to the management and systems development necessary to implementing a change in the way an organisation works. Adopting an endogenous approach is a major organisational as well as policy shift. Linn et al. (2010, 45) note that:

> "ultimately scaling up is about the values and mindset of the people engaged in development and development assistance ... More importantly, IFAD's Board and managers would need to sign on to a fundamental shift in mindset and orientation and share this with all of IFAD's staff – and ultimately all of IFAD's partners on the ground – so that for every rural poverty intervention that IFAD supports two questions are asked as standard practice: 'Is this intervention scaling up our prior experience and/or that of others?' and 'If this interventions works, should it and how could it be scaled up?'"

Implementing a new approach or paradigm requires a new mindset across an organisation. To achieve that mindset, systems, processes, structures, motivation, and rewards need to be in place.

Conclusion

For a Northerner, there is a lingering concern about *endogenous development* and its susceptibility to focus on the local at the cost of ignoring the big picture. It is too easy to focus on the local, case by case. Many of the contributions to this volume centre on the local and everyday. But when we talk about development in an era of globalisation, endogenous development advocates also need to think big. Making change solely on a village by village basis, where cultural context varies, is impractical. While we do not need blueprints or formulas for endogenous development, we do need a set of values and a general framework. George Ayittey calls for a focus on building the basic institutions of independent judiciary, a competent and independent civil service, a free and independent media, or a strong and independent central bank: "[t]he establishment of these institutions would empower Africans to instigate change from within" (http://blog.ted.com/2009/04/09/ayittey_on_dead_aid/). More and more Africans are thinking big, and their voices are increasingly heard through new media and in the North. They succeed because they can speak the language of the North while retaining their African roots.

My own experiences as a development practitioner in Bangladesh, China, and elsewhere (including my own country) convince me of the necessity of endogenous leadership. With a weak state, Bangladesh NGOs have emerged from the grassroots to have an impact on reduction of poverty and inequality far greater than that of government. They did it through an understanding of Bangladeshi communities, through experiential learning and willingness to test, perhaps fail, learn, and achieve. They used donor assistance and advice, but they were not driven by donor requirements. China achieved perhaps the greatest reduction in poverty in human history by controlling their own approach to opening up to the market economy (post-1978), and not by following the donor neoliberal requirements. Where endogenous leadership emerges in Africa, at whatever level, donors have a great role to play in making space for it. Endogenously-led development in Africa will succeed because of ethical and talented African leadership, from the grassroots to the national and global levels.

Notes

1. The Washington Consensus refers to a set of policies around macroeconomic stabilisation and structural adjustments imposed on borrowing countries by the IMF and the World Bank in order to achieve economic growth. The original 10 policy prescriptions focused on stability, opening of economies, and property rights. In general, it is thought that the policies failed to deliver promised growth and that they are unrealistic for poor countries (see, for instance, Rodrik 2002).
2. Easterly (2002) points out that large donor investments in many developing countries failed to produce growth.
3. Rodrik uses the example of China. He also notes that the Washington Consensus is more a description of what developed economies look like than a roadmap to getting there.

References

Adelman, I. 2001. "Fallacies in Development Theory and their Implications for Policy." In *Frontiers of Development Economics: The Future in Perspective*, edited by G. M. Meier, and J. E. Stiglitz, 103–134. New York, NY: Oxford University Press.

Anderson, M., D. Brown, and I. Jean. 2012. *Time to Listen: Hearing People on the Receiving End of International Aid*. Cambridge: CDA Collaborative. Accessed May 29 2014. www.cdacollaborative.org/media/60478/Time-to-Listen-Book.pdf

Ayittey, G. 2009. "On Dead Aid." TED blog. Accessed May 29, 2014. http://blog.ted.com/2009/04/09/ayittey_on_dead_aid/

Chege, M. 2008. "Kenya: Back from the Brink." *Journal of Democracy* 19 (4): 125–139.

Collier, P. 2007. *The Bottom Billion: Why the Poorest Countries are Failing and what can be Done about it*. Oxford: Oxford University Press.

Diamond, L., and J. Mosbacher. 2013. "Petroleum to the People: Africa's Coming Resource Curse – and how to Avoid it." *Foreign Affairs*. September – October 2013. Accessed July 24, 2014. http://www.foreignaffairs.com/articles/139647/larry-diamond-and-jack-mosbacher/petroleum-to-the-peopledictionary.reference.com/browse/endogenous

Easterly, W. 2002. *The Elusive Quest for Growth*. Cambridge, MA: MIT Press.

Escobar, A. 2004. "Imagining a Post-development Era." In *Anthropology of Development and Globalization*, edited by M. Edelman, and A. Haugerud, 342–351. Oxford: Blackwell Publishing.

Fukuyama, F. 2004. *State Building: Governance and the World order in the 21st Century*. Ithaca, NY: Cornell University Press.

Harris, G. 2013. "Rival Economists in Public Battle over Cure for India's Poverty." *New York Times*. Thursday, 22 August 2013.

Holcombe, S. 1995. *Managing to Empower*. London: Zed Books.

Linn, J. F., A. Hartmann, H. Kharas, R. Kohl, and B. Massler. 2010. "Scaling up the Fight Against Rural Poverty: An Institutional Review of IFAD's Approach." Global Economy & Development Working Paper 43. Washington, DC: The Brookings Institution.

Moyo, D. 2009. *Dead Aid*. New York, NY: Farrar, Strauss and Giroux.

OECD. 2005. *The Paris Declaration*. Organisation for Economic Cooperation and Development, Development Assistance Committee. Accessed May 29, 2014. www.oecd.org/dac/effectiveness/34428351.pdf

Rodrik, D. 2002. "After Neoliberalism What?" In After Neoliberalism: Economic Policies that Work for the Poor. A Collection of Papers Presented at a Conference on Alternatives to Neoliberalism, Washington, DC: The New Rules for Global Finance Coalition, 9–20.

Rodrik, D. 2006. "Goodbye Washington Consensus, Hello Washington Confusion." *Journal of Economic Literature* 44 (December 2006): 973–987.

Sen, A. 1999. *Development as Freedom*. New York, NY: Knopf.

UNDP. 1990. *Human Development Report*. Oxford: Oxford University Press.

Van de Walle, N. 2001. *African Economies and the Politics of Permanent Crisis*. Cambridge: Cambridge University Press.

Weaver, J. 2002. "The Problem with US Development Policy." Department of Economics, American University; draft communication for comments.

World Bank. 2000. *Voices of the Poor*. New York, NY: The World Bank.

Indigenous languages and Africa's development dilemma

Mariama Khan

Most African states like The Gambia use European languages for state activities and formal education. Africa has been a global pilot site for "transplanted" development initiatives with apparently consistent outcomes: failure, medium triumph, or unsustainable "success stories". Its natural resources have been fully exploited, perhaps at the expense of resources like mother-tongue languages. Sidelining mother-tongue languages as the medium for the translation of the voice of the state, explains the gap in cultural relevance of many borrowed development initiatives, but also the neglect of workable endogenous practices. Africa must look inwards and exploit its indigenous language assets to benefit sustained development.

La plupart des États africains comme la Gambie utilisent des langues européennes pour les activités publiques et l'éducation formelle. L'Afrique a été un site pilote mondial pour des initiatives de développement « transplantées », avec des résultats apparemment constants : échec, triomphe moyen ou réussites non durables. Ses ressources naturelles ont été entièrement exploitées, parfois aux dépens de ressources comme les langues maternelles. Le fait d'avoir mis sur la touche les langues maternelles comme moyen de traduire la voix de l'État explique le manque de pertinence culturelle de nombreuses initiatives de développement, mais aussi l'abandon de certaines pratiques endogènes réalisables. L'Afrique doit se tourner vers elle-même et exploiter ses atouts linguistiques autochtones pour favoriser un développement soutenu.

Al igual que Gambia, para sus actividades y para la educación formal la mayoría de los Estados africanos utiliza los idiomas europeos. A nivel mundial, África se convirtió en un sitio para la realización de ensayos en torno a ciertas iniciativas de desarrollo "trasplantadas", cuyos resultados aparentemente han sido constantes y repetitivos: fracaso, éxitos a medias o «historias de éxito» insostenible. A pesar de ello, los recursos naturales de este continente han sido totalmente explotados, quizás a costa de ciertos recursos como los idiomas autóctonos. El menosprecio mostrado hacia los mismos como medios para expresar la voz del Estado explica la existencia de una brecha en relación a la pertinencia cultural de muchas iniciativas de desarrollo trasplantadas y a la desestimación concedida a las prácticas endógenas viables. Por lo que, para impulsar su desarrollo sostenible África deberá mirar hacia dentro y comenzar a utilizar sus idiomas autóctonos.

Introduction

This paper revisits the question, what development barriers do African countries face as a result of the sole use of a foreign language in governance and education? Using The Gambia as a case study, the paper further explores what the development benefits of endogenous languages in Africa are. It focuses on literacy, cognitive development, learning achievements, governance, access to information, and accountability as fruits realisable from having " ... *an authentic means of national expression*" (Wauthier 1978, 38).

The debate on valorising African languages by no means indicates hostile disposition towards European and other foreign languages, nor places blame on Africa's colonial history, but rather is a progressive discourse intended to redirect the current African linguistic dilemma to a more rewarding turn. As Gandhi noted:

> "Is it not a painful thing that, if I want to go to a court of justice, I must employ the English language as a medium, that when I become a barrister, I may not speak my mother-tongue and that someone else should have to translate to me in my own tongue? Is not this absolutely absurd? Is it not a sign of slavery? Am I to blame the English for it or myself? ... " (Gandhi, quoted in Wauthier 1978, 1–2)

In the spirit of Gandhi's assertion, it is now time to begin the talk on valorising African languages. For the launching of the African renaissance, it is time to work the talk through a steadfast policy of language accommodation and expansion in continental, regional, and national development interventions.

This discourse on African national languages, democracy, development, and cultural relevance begins with an introduction, preceded by the abstract which gives a brief outlook of the continent's development – past to current. This is followed by the literature review and the arguments of the paper. A case study on The Gambia then gives a picture of in-country current scenarios in the national language agenda. The fourth part details examples of endogenous practices and their potential for sustained political and socio-economic development. Recommendations on what needs to be done to achieve the national language vision follow. This also evaluates The Gambia's failures in progressing with its indigenous language agenda since independence.

Trapped by colonial lyrics in a post-colonial dance

Political independence in former colonial territories in Africa did not translate to the erasure of the colonial footprints. Institutional structures and educational systems inherited at independence remain the platforms for governance and education. African political elites are locked unconsciously into outdated colonial structures of divide and rule (Khan 2009). In statecraft, this adherence to the colonial past has weakened the development of modern institutions in Africa. Consequently, the African state is designed in a way that ideas of the state, power, and development are superficial coatings. The lack of a deeper socio-cultural foundation based on national languages denies the nation the enriching effects language accommodation could yield in the creation of modern state institutions. As May (2003, 128) observes, in the process of state making, the national language is "*a form of practice, historically contingent and socially embedded*". Therefore, broadening the linguistic base of the state strengthens institutions, and legitimises power, authority, and the national vision. Jeremy Waldrons (cited in May 2003, 151) states that, fundamentally in liberalism, "*a social and political order is illegitimate unless it is rooted in the consent of all those who have to live under it*". That consent is expressed through the societal voice that emanates from languages local to the context. It is only through such languages that the systemic cycle of inequalities, exclusions, low accountability, and a huge language gap that chokes development initiatives in most African countries can be avoided.

Development must be grounded in people's realities, culture, needs, and ways of life. The right linguistic foundation and a sound cultural footing nourish the national image and aspirations. The aspirations of a nation are better expressed and pursued through the indigenous language.

This can guarantee that development initiatives will have better outcomes. Hence, the use of a foreign language, especially at the expense of mother-tongue languages, removes the cultural ethos in the expression of the state. On the importance of language in defining people's aspirations, Fishman (2002, 81) argues that in taking the language out of a people's culture, you take *"away its greetings, its curses, its praises, its laws, its literature, its songs, its riddles, its proverbs, its cures, its wisdom, its prayers"* – and, I would add, its strength, hopes, and dreams. Language rationalises the entire gamut of social, political, cultural, and religious understanding of a given society. It forges links between different social entities. Consequently, it shapes their realities based on inherent sociological references shared by the language community.

As such, for rulers and the ruled to follow uniform sets of realities that shape societal aspirations, linguistic democracy must prevail. For countries to realise the social, cultural, and economic potentials, citizens need to be able to access both their traditions and modern information and opportunities through the mother tongue. In addition to this, for countries to better integrate into the global world, there is a need to speak languages of wider communication. Language is capital, as Frantz Fanon (1963) has argued; the more languages one speaks and writes, the better resourced that person is.

Malian anthropologist Pascal Baba Couloubally (2006, 25) argues that, in Africa, *"the problem of the State is one that concerns the language of expression and exercise of the state"*. In Mali, for example, two out of 10 Malians are literate in the official language, French. Low literacy rates in French obstruct mass understanding of the state's aspirations. In Mali, as in The Gambia, the use of the foreign language hinders the development of a shared vision between the state and people. Limits on access to information and communication barriers arising from the language gap stall participatory democracy and accountability.

Sceptics argue that the use of foreign languages, such as English in The Gambia, is a better option for African countries given the multiplicity of languages. The foreign language is seen as a "neutral language". Since it is not associated with any particular group in the country, it fosters national unity. But as Renan (cited in May 2003, 41) argues, *"Language may invite us to unite but it does not compel us to do so"*. This flawed assumption of the neutral status of the foreign language in Africa ignores, first, that Africa has a tradition of language shifting and assimilation. Generally, people are multi-lingual and *lingua francas* exist in societies as socially accepted languages for inter-tribe communication. Secondly, it ignores the reality that there is no neutral language. European languages brought to Africa are not just words but words imbedded with ideologies, at the heart of which was a "civilising mission". Thus, there are enormous implications for African identity if it is expressed only through the European language. On their ability to foster unity among Africans, it brings to mind a question a former professor once asked me: *"Do African people need a European language to unite them?"* The evidence of wars, secessions, and conflicts in Africa suggests that European languages have not united Africa, nor lessened conflicts. Additionally, despite the multiplicity of languages in Africa, certain languages evolve in society, as the language of wider communication and the *lingua franca*. Such a language is spoken by all irrespective of their ethnic and linguistic background. There are more than 10 languages in Senegal but all Senegalese speak Wolof irrespective of their tribal affiliation. In Mali too, almost all Malians speak Bambara, as one of the language groups in the country. In Northern Nigeria too, almost everyone speaks Hausa. And certainly Hausa speakers are not the only people living in the region. All these examples show that African societies have in various ways negotiated the linguistic terrain to uphold a common language that everyone can associate with. Thus, language multiplicity enriches society and it is by no means a disadvantage to the country. In fact, using a European language in an Africa country contributes one more language to a landscape of multiple languages.

Restricting indigenous language use in the state and educational settings clearly has undesirable effects on the attainment of African development goals. The path of trial and error that has become the routine approach in African development interventions could tread on assured tracks if indigenous languages are mainstreamed in national development efforts. So far, indigenous languages obtain a lacklustre attention in testing solutions to Africa's elusive development. Consequently, the continent continues to pay the price for this.

Speaking European languages, as I have argued earlier, does have rewards since language is a resource. However, in a situation where all the laws, the education, and the business of a country are conducted in a foreign language understood by a minority of the people, the ruled and the rulers face distinct realities. Participatory democracy is reduced to the rituals of elections, and political accountability is lost in the façade of linguistic inequalities. Language mediates behaviour. It has a determinable and causal relation to it. "*A language long associated with the culture is best able to express most easily, most exactly, most richly, and with more appropriate overtones, the concerns, artefacts, values, and interests of that culture*" (Fishman 2002, 81). Language functions as an "*instrument of communication and rational thoughts*" (Mazrui and Mazrui 1998, 62), and a tool for social liberation. McLaughlin (2001) suggests that in the 1980s–1990s amid widespread economic difficulties, Dakarois adopted written forms of Wolof, which was largely an oral language, to legitimise their urban identity. Wolof became a political instrument for social change and democratic transition in Senegal. The lack of linguistic democracy in a country denies the people social and political freedoms. Terwilliger (1968) argues that the nature of a person's language determines his entire way of life, his thoughts and other forms of mental activity: "*Not only is language a device for communicating our ideas and intentions to others, but also a device for regulating our own behaviour*" (1968, 9).

The language gap caused by the mandatory use of the foreign language for most functions, at the expense of better understood mother-tongue languages, creates an unequal power dynamic. "*Language is about power and control*" (Khan 2009, 16). Low foreign language literacy levels fail to support full participation in policy-making and other national development processes. Mowarin and Tonukari (2010) reinforce my argument that government intervention is necessary to rectify the current inadequacies in the use of both the foreign and mother-tongue languages. Whether from large-scale or micro-level language planning approaches (Baldauf and Kaplan 2006), the state has key roles to play in redeeming African languages. Social exclusion, information and communication barriers, and low literacy promote a poorly defined national vision and retard opportunities for national consensus building in society. The elites swiftly free-ride and engage in unaccountable governance.

Literacy can be used for colonisation (Jenkins 1991). But also as Cummins (1995, 89) argues, literacy can be explicitly focused on issues of power as seen in the work of Paulo Freire, who highlights the potential of written language as a tool that encourages people to analyse the division of power and resources in their society and to work toward transforming discriminatory structures (cited in Burnaby and Reyhner 2002, 2).

Competence in a language affords access to information. It helps the individual and organisations to scrutinise power and hold it to account. Hence, the right to education should be complemented by the right to be taught in one's language. Balduaf and Kaplan (2006) argue that current language policies in most African countries are driven by market forces, justifying the continuation of colonial trends in educational planning and policies. But even where the language of trade is different from the official language, as in Guinea Bissau and Niger (Hovens 2003), "*bilingual education is a cognitive advantage and not a deficit ...* " (May 2003, 144). Yuka and Omoregbe (2010) demonstrate that even minority African languages, among them Edo in Nigeria, have the requisite linguistic structures to accommodate scientific and non-scientific knowledge production. The benefits of mother-tongue education can only be accrued by securing an enlarged role for it in national development. It has to be the words of the state, the understanding of the citizens, and the call of the private sector. This will heighten national esteem, ensure higher learning achievements and better cognitive development, and enhance the creativity needed for technological innovation and socio-economic progress.

Studies show that in bilingual education and bi-literacy situations in different countries, drop out and repetition rates are very low. This results from the benefits of incorporating the mother tongue as the language of instruction in the school system. In Guatemala 40% of Maya-speaking populations enrolled in schools without mother-tongue education drop out, but this changed with the introduction of mother-tongue literacy in some schools. A study in Senegal found that an agricultural extension worker understands better the insects that destroy crops in a Pulaar dictionary than through an agricultural extension booklet in French. Of the 62% of children enrolled in French language schools in Senegal, 75% drop out by the seventh grade as a result of failing (Hutchison 2006). In Tanzania, the Philippines and India, mother-tongue education increased student-teacher and parent-teacher communication, improved performance, self-esteem, and cognitive skills ensuring independent thinking and creativity which are important for development (Benson 2004). A study of 1,664 children in 131 classrooms and in 163 observed lessons and 720 interviewed parents showed pupils who study first in the mother tongue perform better in acquisition of knowledge on substances, liquids, weights, and measures. They did better in math than mono-lingual (French or Portuguese) studying pupils (Hovens 2003). Similarly, in bilingual education where the mother tongue is one of the languages of instruction, students have better reading and writing skills. For example, in Mali, 20% of pupils aged 6–9 read four languages. Other evidence comes from Nigeria where 103 schools teach four languages (30% of schools) (Woolman 2001). Sokoto and Yoruba Land (Nigeria) also have models of Freirian liberation literacy based on use of their indigenous languages (Omoniye 2003). These findings suggest that mother-tongue education promotes language acquisition and higher performance, equipping students with tools to engage with livelihoods and governance.

Clearly, mother tongue languages are an important asset for Africa's regeneration. From the case studies cited, it has unquestionable benefits for mass literacy, creativity, and innovation. Neglect of African languages as a valid medium for the transmission of knowledge is denying the global village resourceful insights into African knowledge. It also limits African scholarship and intellectual achievements while disconnecting links between citizens and leaders in society. Below, I explore how The Gambia was forced to reconsider education in the mother tongue following failures in education.

The Gambia's indigenous language trials

From 1970 onwards, education in The Gambia deteriorated. Low literacy rates and low education quality led to low national manpower skills development and poor political awareness. In 1965, 35% of teachers in the country were unqualified. By 1980, this number rose to 66%. Over four decades, policies promoting mother-tongue education received lip service, but there has been no serious effort to implement them. The 1997 constitution provides in Chapter 111 of subsection 30 "*All persons shall have the right to equal educational opportunities and facilities ... *". Subsection 32 states: "*Every person shall be entitled to enjoy, practice, profess, maintain and promote any culture, language, tradition or religion subject to the terms of this Constitution ...* " Constitutional guarantees are reflected in stated policies. The educational policy for 1988–2003 called for the introduction of national languages in schools. The subsequent policy period 2004–15 has the same provisions. Both policy periods have seen no implementation of the policy intent.

However, serious thought on the policy statements on national languages was forced in 2007 when a literacy test conducted by the Ministry of Education showed stunningly poor results. In the survey conducted by the Ministry (see Mission Report, June-July 2012 and Report on Read Gambia September 2012), 120 randomly selected third and second graders from 40 schools were tested on pre-reading and reading skills, comprehension, spelling, and punctuation skills. Results showed only 32% of third graders had correct letter identification skills, 34% had correct phoneme pronunciation, and 46% could not read a single word or connected text. Reading levels were 10 words per minute, but there was no comprehension of the material read. Fifty-four per cent of second graders could not read a single word per minute.

This uncovered reading gap subsequently led to a pilot introduction of five national languages – Mandinka, Olof, Pulaar, Jolaa, and Saranhuley – on an assumption (and evidence from other countries) that children learn faster in their own language. Reading and writing skills acquired in the mother tongue can be effectively used to learn a second language. One hundred and twenty-five classes of 30 pupils each in the country's six educational regions were targeted. New materials on national languages were developed: orthographies for each language, primers, readers, and scripted lessons for uniform teaching. Additionally pedagogic support through coaches was provided, and over 3,000 teachers were trained. For programme evaluation purposes, some coaches were also trained on observation and data collection.

Project challenges and initial outcomes

Classes for the national language programme were scheduled for daily one-hour lessons. Punctuality challenges from both pupils and teachers led to some schools not completing the curriculum. Teachers showed mixed levels of motivation. Pilot schools surveyed at baseline were selected for a post-test evaluation, comprising 1,246 pupils, with 475 in a comparison group. The survey tracked between 23%–57% pupils at baseline. Of the 25 programme classes surveyed, each language had 14–15 lessons. After 15–20 weeks of scripted lessons, programme objectives aimed to have 85% of the pupils able to name 80% of the letters, read at least one word in one minute in a connected text, and all pupils showing progress in their reading by the end of the pilot.

Pupils from both pilot and comparison groups were asked to read the 26 English letters of the alphabet in random order. A post-test was added with children asked to read, in addition to the English letters, the special characters/orthographies in each of the national languages. For the reading of syllables part, the pupils were given 20 syllables to read out. In the reading of words sample text, an enumerator pointed out 20 commonly used words which the pupils were asked to read. On connected text and reading comprehension, pupils had one minute to read a sheet with paragraphs in the national language (letter size 36). Also three reading comprehension questions were set on passages with 23–25 words at baseline and about 45 words at posttest (the Sarahully language had 49 words). A previous 2009 Early Grade Reading Assessment (EGRA) text on connected text and reading comprehension questions in English was used as reference. Pupils had one minute to read a passage and five comprehension questions. The findings of the above study showed the strengths and weakness of the programme. Overall it demonstrates that mother-tongue education has benefits that can be lost by sticking to the sole use of the foreign language in schools.

While differences in individual school achievements are not accounted for in these tables, there are significant achievement differences between pilot and non-treatment groups. Pilot schools generally performed better than the control schools in all the tasks. Even though programme goals are not met, these initial results point to the importance of indigenous languages in ensuring higher learning achievements and better cognitive development in children (see Tables 1 and 2).

Table 1.   Percentage of students reading at least 80% of letters correctly.

	Baseline		Post-test	
	Comparison	Pilot	Comparison	Pilot
Jola	8.0%	15.9%	31.1%	49.1%
Mandinka	10.8%	10.7%	3.2%	36.9%
Pulaar	16.5%	7.6%	29.5%	68.7%
Sarahulle	5.3% ($N=3$)	2.0% ($N=3$)	16.5%	41.1%
Wollof	6.7%	16.7%	12.6%	56.8%

Source: The Gambia Ministry of Education, 2011.

Table 2.   Percentage of children reading at least one word in the national language passage (45-49 words, one minute).

	Baseline		Post-test	
	Comparison	Pilot	Comparison	Pilot
Jola	0%	0%	3.4% ($N=6$)	42.4%
Mandinka	1.4% ($N=1$)	0%	3.2% ($N=1$)	50.0%
Pulaar	1.4% ($N=2$)	0.7% ($N=3$)	2.1% ($N=3$)	44.3%
Sarahulle	1.8% ($N=2$)	2.0% ($N=3$)	4.3% ($N=4$)	41.5%
Wollof	0.7% ($N=1$)	0.5% ($N=1$)	43.3%	31.6%

Source: The Gambia Ministry of Education, 2011.

Indigenous languages, endogenous development practices, and African renaissance

Beside the payoffs in cognitive development, better learning achievements, and securing mass literacy, mother-tongue languages can serve as ideal springboards for the national adoption of useful endogenous practices. There are many culturally sensitive, organic practices that exist in Africa. Their incorporation into change interventions can strengthen modern democratic and development thought. However, this incorporation can only be done when indigenous languages are made the language of the state and education. This is because endogenous practices are embedded in the language and as such the use of the language and the creative adoption of the practice(s) form a unified continuum. Arguably, in Africa, the lack of sustainability in positive development interventions or governance changes is partly due to such interventions being rooted in ideologies alien to the indigenous expression. On the other hand, differences in the language of the state and the language of society account for the neglect of useful endogenous practices or the creative adoption of such practices as the foundation for some modern institutions of the state.

Recently, for example, while on a visit to Dakar and talking on values and accountability, I learnt that in Fouta Toro (western Senegal), home to the Pulaar-speaking Tukulor ethnic group in Senegambia, there exists an organic democratic Imam selection process. As a value-oriented tribe, the Tukulors are known for their discrete etiquette, strong Islamic identity and social protocols on character, truth, justice, cordiality, and good personal behaviour.

A proposed candidate for Imamship has to have, first, unquestionable erudition in Islam. He must also be a man of dignity with the highest moral reference. Securing the nomination of fellow men is followed by a community event for a final endorsement by women. If a candidate fails this, he is disqualified for the Imamship. This process is based on the belief that women have a better awareness about the true moral character of a man than his peers; hence they possess the ability to know the hidden sides of the male.

Transposing these ideas to our current times, the gendered nature of assuring accountability is not just anthropologically interesting, but it resonates with the ways in which respected personalities in military, politics, finance, and governance were pulled down from grace through women who knew what others did not know about their character. Think of the former IMF Executive Director Dominique Strauss-Kahn, the American General David Patraeus, or former Prime Minister Silvio Berlusconi of Italy. In developing modern systems of accountability, Africans can draw on their own cultural traditions to check abuses of power.

Another example of a worthy endogenous practice is the *Bulunda*, a restorative justice forum that is presided over by the elders of the founding families of Brikama, the regional capital of The Gambia's western region. Described as a forum of truth and justice by both accused and plaintiffs, conflicts are negotiated in the *Bulunda* behind closed doors, with only elders, victims, and offenders in attendance. Cultural values and norms of justice rooted in the Mandinka traditions guide their deliberations. It has a 99% rate of satisfactory mediation of cases brought before it. The ability to safeguard interpersonal relationships, community order, and social peace, without jeopardising the fundamentals of social justice, earns the *Bulunda* its long-standing unquestionable reputation as the forum of truth and justice. Understandably, it is more trusted than the Alternative Dispute Resolution (ADR) Programme which is situated in the context of the formal justice system of the country. This is because the people can better rationalise and identify with the *Bulunda*'s justice system. Its ideology resonates with their ideas of justice, truth, and needs, as opposed to what the ADR programme and its linguistic alienation represents for the people. Adopting the *Bulunda* techniques of justice delivery could creatively bolster the rule of law in the country. But this will only be possible if the right linguistic bases for it are available.

Beyond rhetoric to action in mother-tongue language valorisation

Undoubtedly, mother-tongue education will stimulate creativity and innovation in the African development and democratic agenda. However, realising its benefits requires strategy, policy, planning, and consistent commitment. This necessary shift will also consolidate the gains in the existing languages of wider communication (e.g., English) while avoiding the barriers they pose to African development. Mackey (2010, 19) rightly states: "*language policies depend on those who make them. States do not make policies; people do ...*" And their actions depend on their perception of the present and vision for the future. Policymakers and implementers can make or unmake the language valorisation agenda in Africa. If there is political will to harness benefits in the indigenous languages, Government must guarantee that mother-tongue education is *top priority in its political, economic, and education* agendas. Any gains from mother-tongue education will rest on the level of national political will. Government activism for the valorisation of indigenous languages deserves the support of civil society and other development partners. Baker (1992, 9) notes: "*If a community is grossly unfavourable to bilingual education or the imposition of a 'common' national language ... , language policy implementation is unlikely to be successful.*" A form of mental reengineering is necessary for society to have the right disposition towards this language valorisation agenda. As Asante (2007, 72) notes, for the regeneration of Africa, "*Where there have been others, our thoughts must now become our own thoughts. Where we have forgotten our own traditions of knowledge and information we must remember*". Through this, the birth of the long-talked of African Renaissance will be witnessed. But again, success or failure will depend on social attitudes. "*Attempting language shift by language planning, language policy making and the provision of human and material resources can all come to nothing if attitudes are not favourable to change*" (Baker 1992, 21). This demands investing heavily in the agenda and moving from mere rhetoric to concrete actions. Ministries of education should have functional language planning units to implement language policy in collaboration with other stakeholders. This should have an across-the-board focus, with primary, secondary, and tertiary education as its target.

Creating jobs in mother-tongue languages in all aspects of the economy should be the goal. Research and publications on languages should be encouraged. As in all modern languages, mother-tongue languages should be kept alive through the development of new terminologies; if these languages can be studied by Western scholars for their linguistic forms, values and system, it means they are like all other languages (Ladefoged 1968). Scholarship in the mother-tongue languages can be promoted by creating a cultural space that makes the mother-tongue language competitive and economically viable. The language should be a school subject and the medium for learning geography, archaeology, history, communication, and politics: "*Creating cultural space is very important for a language if it is to become competitive within its own culture*" (Fishman 2002, 89). Teaching excellence should be the target, with teachers motivated to develop excellence in their skills.

The state should express itself using mother-tongue languages. People should be able to access information through the indigenous language, whether it is in the legal system or other sectors. People's birth certificates and other national documents should have information in the national language. The Canadian example where English and French "*shared so fully the burdens of a modern industrial state*" (Mackey 2010, 57) should be very instructive on this.

Indigenous languages can facilitate the regional integration agenda in Africa. This requires clear language policies at both national and regional levels. For regional integration, languages with wider communication presence such as Pulaar, Hausa, and Mandinka in West Africa, and given the existence of written forms for these, could be used as stepping stones.

In addition, both the print and non-print media are useful tools for language valorisation. Already, in most African countries, mother-tongue languages are used in broadcasting. This should be extended to a print media that writes in the local languages. These languages should also be used in rituals and other ceremonial events; assembly debates with story and poetry competitions in the indigenous languages encouraged at all levels in society. African elites should speak their languages in the home instead of French, English, or Portuguese. For some elites, speaking European languages at home is a status symbol. However, by failing to speak their language at home, they send negative signals to their children about the desirability of the indigenous language. This hurts the language and denies it social esteem.

The history of educational development and mother-tongue languages in The Gambia is a story of how good intentions can become missed opportunities and failures. In the spirit of learning from the past, I briefly examine how Gambian policy implementation foundered. Using selected newspaper articles from 1979, which was a landmark year as the United Nation's International Year of the Child, I highlight what went wrong with policy implementation and the genesis of the current literary crisis in the country.

The Gambia's missed opportunities

The Gambia has long expressed a supportive policy towards mother-tongue languages. In the five-year development plan that was under implementation in 1979, it promised to develop Mandinka, Wollof, and Fula to national recognition. Dembo Jatta, Gambia's then education, youth, and sports minister said during the 1979 inauguration ceremony of the newly established Advisory Council for Education. He noted that the curriculum inherited from the colonial administration was not responsive to the needs of the Gambian economy. Hence, the goal for 1979/

1986 Educational Policy was meaningful education. However, the divergence of rhetoric and action was apparent in the numbers of unqualified teachers, which rose from 35% in 1965 to 66% in 1980 (See, *The Gambia News Bulletin*, August 19, 1980).

The Gambia's government adopted some endogenous practices that served its purposes but neglected others. A selective and symbolic support of endogenous practices thrived from 1959, when the People's Progressive Party (PPP) adopted the hoe and the axe as party symbols to harness support from the mass agricultural population in the provinces, until 1994 when it was overthrown in a coup. Voted into office by provincial votes in 1965, the Jawara government's development philosophy was *Te Sito*, a Mandinka word and a local concept derived from farming.[1] Literally, the word *Te* means waist, *Sito*, means tying. It meant self-help for self-improvement or the "*the act of lifting up oneself by one's own bootstraps*". As a term, it embodies both socialist and capitalist notions of work and progress. *Te sito* had deep roots in Gambian community orientation and social organisation. The common good is pursued through community mobilisation, whether it is through voluntary labour or in-kind or monetary contributions towards a cause. For example, in January 1979, as the world recovered from the recession caused by the oil crisis, 200 PPP youth in Gambiasara, Upper River Division, harvested 120 bags of cotton from their communal farm. They also conducted a general cleaning of their village (See, *The Gambia News Bulletin*, January 11, 1979). Community-based efforts like these have provided social insurance to members in the face of global economic hardships and their localised echoes in the national and local fronts. Organising for the general interest of members has always been part of community social organisations. The age grade system, which is derived from communal social structures (Renner 1985; Hughes and Perfect 2006; Senghor 2008), has been one of the mobilising mechanisms in trying to address low-key "*participation of urban youth in national politics ...* ", in what was nation-building for youth leaders (*The Gambia News Bulletin*, January 6, 1979). As a development philosophy, its potential was boundless in helping to shape a shared national vision in the country. However, when structural adjustment came, *Te Sito* died.

Jawara came to power at a time when communal social structures and modalities of social cooperation for the common good were a solid part of the social fabric. A vast fertile ground for building, creatively, culture-relevant modern national institutions out of existing practices prevailed. However, his regime used endogenous practices well when they served the regime's agenda and they neglected these traditional values where they might contribute to social consciousness and long-term political awareness. It is this selective use of traditional practices by African politicians to serve political expediencies that is problematic to the endogenous language development agenda. In fact, even the post-Jawara era has seen the mutative adherence to the notions of traditional community activism through the age grade system. But again, use of endogenous practices was designed to entrench and protect the interests of a select few in the country. The formation of the "Green Boys" political group after the 1994 coup took its inspiration from the age grade system. The age grade system epitomised a noble and just oriented community philosophy in social structures, but unfortunately, the Green Boys became a corrupt adoption of all the good in that form of social organisation. The Green Boys became a bully to political opponents of the ruling party and its interests. As a result of their bullying acts, it can be argued that the Green Boys did what age grade politics in community life and in its untainted philosophy would have deeply frowned upon. The formation of the group was, in a negative way, an act of borrowing from worthy social practices. As political bullies, the Green Boys stepped out of the bounds of social morality and took on an image of their own. But this is an example of how the selective prejudice of politicians in adopting endogenous practices can be misguided, too. Similarly, a lack of direction can also beset the adoption of such practices. For instance, in terms of teaching national history in schools, one can recall initiatives like the collection of oral and written *tarikkhas* (accounts) of the Kingdom of Kaabu, including migrations of the major families of Manding to Kaabu, their settlement, relationship with local people, internal migrations, the structure of the empire, and the eventual fall of the empire (See, *The Gambia News Bulletin*, January 13, 1979). This effort could have been treated as prized history in The Gambia. As history it could have been integrated into the school curricula; instead, once it was collected, it was archived and forgotten. An exclusive club of Gambian historians did write a book in English, on Gambian history, but again, the usefulness of this history was questionable as the book had limited outreach in society. In addition to this, no indigenous language was used, to better situate the history, in the writing of the book.

In 1980, the National Literacy Advisory Committee (NLAC) presented the "first" Mandinka dictionary it published to President Jawara. One hundred and fifty-eight pages thick, it was funded by the Rural Development Project, at cost of D3000 and with a production time of 18 months. Before this, there was a previous Mandinka dictionary by Gambianists David Gamble and George O'Halloran, both Western scholars. There was also another Mandinka dictionary in Senegal which had minimal differences with the one produced by NLAC. Having invested in the dictionary, the government did not develop a plan to distribute the dictionary or to encourage its use in ways that would support mother-tongue education. The NLAC's distribution plan for the Mandinka dictionary was to put one copy in each government office and school. They did not address key questions necessary to support the use of the dictionary, for example: what can be done with that one copy? Who would use it and for what purpose? What steps needed to be taken to create an environment conducive to mass voluntary interest in written Mandinka? Was there any easy access to other reading materials, including textbooks that would make the diction- ary a more useful reference material? NLAC did have 16 schools in the provinces where night classes were conducted in Mandinka (See *The Gambia News Bulletin*, August 9, 1980). Another group, the World Evangelist Council (WEC), was also engaged in the teaching of Man- dinka upcountry. For NLAC, those informal night classes were considered adequate implemen- tation and constitute the sum of government plans to develop Mandinka, Wollof, and Fula to national recognition.

This example illustrates the gap between policy and action and the inability to develop a serious plan to implement a major language and education innovation. There are other lingering questions, for example, about why multiple dictionaries were produced in Senegal and The Gambia. The two countries could have collaborated and partnered on this project and through that, implemented a mass literacy programme in Mandinka in both countries, benefitting national literacy as well as regional communication. However, there was a lack of informed strategy in the Gambian govern- ment's articulated policy in its pursuit of the valorisation of indigenous languages. The same can be said of the Senegalese attempts geared in the same direction. Building synergies across all levels, inter-state and at the regional level, could bring vast resources to the language valorisation agenda. In turn, these will have positive effects on democracy and development in Africa. The crea- tive adoption of endogenous practices as a basis for modern institution building in Africa can only transpire within a context of language accommodation that will mainstream the use of mother- tongue languages in national development. This is because these practices are better expressed through the language that is rooted in the ideology and sociology of the society.

Conclusion

Clearly, the success of harnessing mother-tongue languages as an important development asset in Africa will rest on genuine political will and conscious planning. From examples cited above, the magic is commitment and dedication from all sectors in the country. The school learning experi- ence should have direct linkages with national administration and the economy. People should learn in their own language, use it to participate in governance, and also derive livelihoods from it. Ensuring freedoms and socio-economic well-being should be the goal of national language valorisation.

Note

1. Sir Dawda Kairaba Jawara was the first President of The Gambia. He ruled the country from 1965 to 1994, when he was toppled in a military coup by the current President Yahya Jammeh.

References

Asante, M. K. 2007. "African Betrayals and African Recovery for a New Future." In Africa *in the 21st Century, Toward a New Future*, edited by Ama Mazama, 71–98. London: Routledge.

Baker, C. 1992. *Attitudes and Language*. Clevedon: Multilingual Matters Ltd.

Baldauf Jr, R. B., and R. B. Kaplan. 2006. "Language Policy and Planning in Botswana, Malawi, Mozambique and South Africa: Some Common Issues." Accessed April 30, 2007. eprint.uq.edu.au/archive/00002935/01/AFRICA_1_LPPAfrica.pdf

Benson, C. 2004. "The Importance of Mother-Tongue Based Schooling for Education Quality." Centre for Centre for Research on Bilingualism. Accessed April 30, 2007. unesco.org/education/en/ev.php-url

Burnaby, B., and J. Reyhner, eds. 1980. "Canadian Tutors to Upgrade Unqualified Teachers." *The Gambia News Bulletin*, August 19.

Burnaby, B., and J. Reyhner, eds. 2002. *Indigenous Languages Across the Community*. Flagstaff: Northern Arizona University.

Couloubaly, P. B. 2006. *Inter-Generational Dialogue and Synergy for the Future. Inter-governmental Forum on Endogenous Governance in West Africa*. Ouagadougou, Burkina Faso: Sahel and West Africa Club/OECD.

Fanon, F. 1963. *The Wretched of the Earth*. New York: Groove Press.

Fiona, M. 2001. "Dakar Wolof and the Configuration of an Urban Identity." *Journal of African Cultural Studies* 14 (2): 153–172.

Fishman, J. 2002. "What Do you Lose When you Lose Your Language." In *Stabilizing Indigenous Language*, edited by Gina Cantoni, 80–91. Flagstaff, Arizona: Northern Arizona University.

Government of the Gambia. 1997. *The Constitution of the Republic of the Gambia, 1997*. Banjul: Government of The Gambia.

Hovens, M. van de. 2003. "Bilingual Education in West Africa: Does it Work?" Accessed March 18, 2007. www.multilingual-matters.net/beb/005/beb0050249.htm

Hughes, A., and D. Perfect. 2006. *The Political History of The Gambia 1816–1994*. Austin: University of Rochester Press.

Hutchison, J. P. 2006. "African Languages: Literature as a Weapon Against African Language Marginalization." Unpublished.

Khan, M. 2009. *Indigenous Languages, The Way to Africa's Renaissance*. Brikama: Sandeng Publishers.

Mackey, W. F. 2010. "The History and Origins of Language Policies in Canada." In *Canadian Language Policies in Comparative Perspective*, edited by Michael A. Morris, 18–66. Montreal: McGill-Queen's University Press.

May, S. 2003. "Misconceiving Minority Language Rights: Implications for Liberal Political Theory." In *Language Rights and Political Theory*, edited by Will Kymlicka and Alan Patten, 123–168. Oxford: Oxford University Press.

Mazrui, A. A., and A. M. Mazrui. 1998. *The Power of Babel, Language and Governance in the African Experience*. Chicago: The University of Chicago Press and Fountain Publishers.

Mowarin, M., and E. Ufuoma Tonukari. 2010. "Language Deficit in English and Lack of Creative Education as Impediments to Nigeria's Breakthrough into the Knowledge Era." *Educational Research and Reviews* 5 (6): 303–308.

Omoniye, T. 1980. "President Receives Mandinka Dictionary." *The Gambia News Bulletin*, August 9.

Omoniye, T. 2003. "Local Policies and Global Forces: Multi-literacy and Africa's Indigenous Languages." *Language Policy* 2 (2): 133–152.

Renner, F. A. 1985. "Ethnic Affinity, Partition and Political Integration in Senegambia." In *Partitioned Africans, Ethnic Relations Across Africa's International Boundaries 1884–1984*, edited by A. I. Asiwaju, 71–85. London Lagos: C. Hurst & Company and University of Lagos Press.

Reyhner, J., G. Cantoni, R. N. St. Clair, and E. Parsons-Yazzie, eds. 1999. *Revitalising Indigenous Languages*. Flagstaff: Northern Arizona University.

Senghor, J. C. 2008. *The Politics of the Senegambian Integration 1958–1994*. Bern: Peter Lang AG International Academic Publishers.

Terwilliger, R. F. 1968. *Meaning and Mind, a Study in the Psychology of Language*. New York: Oxford University Press.

Terwilliger, R. F. 1979. "Urban Youth Action Group." *The Gambia News Bulletin*, January 6.

Wauthier, C. 1978. *The Literature and Thought of Modern Africa*. London: Heinemann.

Woolman, D. C. 2001. "Educational Reconstruction and Post-colonial Curriculum Development: A Comparative Study of Four African Countries." *International Educational Journal* 2 (5): 27–46.

Yuka, N. C., and E. Mercy Omoregbe. 2010. "The Internal Structure of the Edo Verb." *California Linguistic Notes* XXXV (2 spring): 1–19.

Endogenous development going forward: learning and action

Chiku Malunga and Susan H. Holcombe

More than 50 years after independence Africa is yet to move from colonial to post-colonial identity – and to entitlement to determining its own destiny. Increasingly, however, African development thinkers and practitioners are questioning the dominance of externally driven, mostly Western models of development, which they believe have done little to date toward bringing about self-reliant sustainable development. We have observed successful patterns of endogenously led development in East Asia and Brazil. In Africa the papers included here suggest emerging new patterns of local leadership and of resurrecting and renewing cultural and traditional strengths to support modern development. Endogenous development, while a sometimes awkward term, is a concept increasingly informing practice.

Plus de 50 ans après son indépendance, l'Afrique n'est pas encore passée d'une identité coloniale à une identité postcoloniale – et au droit à déterminer son propre destin. Toutefois, de plus en plus, les penseurs et les praticiens du développement africain mettent en cause la dominance des modèles de développement principalement occidentaux, impulsés depuis l'extérieur, dont ils pensent qu'ils n'ont guère fait jusqu'ici pour donner lieu à un développement durable autosuffisant. Nous avons observé des modèles réussis de développement endogène en Asie de l'Est et au Brésil. En Afrique, les articles présentés ici suggèrent de nouveaux schémas émergents de leadership local, et de résurrection et de renouvellement des forces culturelles et traditionnelles pour soutenir le développement moderne. Le développement endogène, même si c'est une expression parfois délicate, désigne un concept qui éclaire de plus en plus les pratiques.

A pesar de que ha pasado más de medio siglo desde que África logró su independencia, el continente sigue sin llevar a cabo el tránsito desde una identidad colonial a una identidad poscolonial y hacia el reconocimiento de que tiene el derecho a determinar su propio destino. Sin embargo, los pensadores y los operadores de desarrollo africanos cuestionan cada vez más el predominio de modelos de desarrollo casi siempre occidentales, que son impulsados desde el exterior. Hasta la fecha, sostienen, éstos han hecho poco en pos del surgimiento de un desarrollo autosuficiente y sostenible. Por otro lado, se constata que en Asia oriental y en Brasil han surgido exitosos patrones de desarrollo endógeno. Los artículos incluidos en este tomo apuntan a considerar la emergencia tanto de nuevos patrones de liderazgo local como de renovadas y resucitadas fortalezas culturales y tradicionales que subyacen al desarrollo moderno de África. Si bien en ocasiones el término «desarrollo endógeno» resulta incómodo, dicho concepto se está incorporando crecientemente en la práctica.

As we promised at the outset of this special issue, there are many definitions of endogenous development, but what unites the definitions is that development to be sustainable must be locally determined and led; development must also evolve organically from local context, culture, and institutions. These are separate but overlapping definitions.

Key learning from this volume:

(1) **Globalisation processes are driving African interest in endogenous approaches, but still limit access to knowledge and voice**.

African thinkers are looking at their own historical and cultural experiences in order to define development goals and as a foundation for building institutions and processes that foster development that is appropriate to local contexts and values (see Malunga, Millar articles). The rapid impacts of globalisation, perhaps even more than the impacts of colonialism, are eroding Africans' links to their cultural heritage and to their history and culture, increasingly limiting the ability of Africans to draw on traditions that serve the future (Banda). Globalisation does this through multiple mechanisms, including consumerism, migration, and new social expectations. Globalisation does create benefits in terms of new ways for Africans to communicate with each other and enables Africans to use new methods of dialogue and problem-solving (wiki – Booker) that resonate with older African traditions. Globalisation does not favour all actors equally. Commercialisation of knowledge and inequalities in access to the Internet and online journals, research funding, and publication in global journals create barriers for African scholars and practitioners to access scholarly information about development and endogenous development. In preparing this volume we co-editors noted that contributors often could not access literature in scholarly publications; this special issue was an attempt to open a door for African thinkers to give voice to their work on endogenous development. Additionally we found that Africans with much to contribute were in such demand from donor agencies and governments that they did not have time to write the articles they would have like to have contributed.

(2) **African development practitioners, sometimes with the support of sympathetic partners, are putting endogenous development approaches into practice**.

Sometimes it is in terms of adapting traditional Rwandan management and accountability systems to modern governance (Rwiyereka) or using traditional Somali practices in community organisations focused on modern goals (Delaney). Combining knowledge of local circumstances with participatory techniques, the development professionals contributing to this issue demonstrate how local knowledge and participation can contribute to resolving conflict over water tariffs (Galaa and Bukari); embed water and sanitation improvements in local traditions (Zakiya); or link small producers to markets (Tinguery). African professionals have the capacity for scientific and social innovation (cassava processing: Fuller-Wimbush and Adebayo); it remains for donors to identify and support African innovation and its scaling up – not just concentrate on innovations originating in the North or moving on to the next innovation without investing in determining its scalability.

(3) **Language remains a challenge for the majority of African states.**

This is particularly the case for those artificially created in Berlin nearly 130 years ago. There are arguments for colonial languages that can be used across multiple ethnic language groups, but the use of colonial languages demonstrably excludes the poorest in most countries and fosters an elitism. It also separates people from traditions that can be expressed most fully in the original tongue. This is an area where creative solutions can both foster learning in mother tongues and access to international dialogue (Khan). For African practitioners and development leaders, the question of language will remain a dilemma. Facility in local language is critical to inclusion and to relying on cultural foundations. At the same time, to engage actively in the global development dialogue, Africans need a global language, but they approach fluency in a global tongue as a challenge of mastering a second or third language.

(4) **Funding agencies have an enormous role to play in fostering and giving space to endogenous development**.

The Paris Agenda and subsequent work of the DAC has given emphasis to country ownership and participation. Country ownership, several articles here suggest, needs to be rooted in African approaches and African leadership. The Paris Declaration, however, sets up a dilemma with its principles. Along with country ownership and a commitment to national capacity building, the emphasis on results measureable in the short term may make it difficult for African governments and civil societies to develop the needed capacities. The donor emphasis on measurable results in the short term creates a perverse incentive for intermediary agencies, governments, and local NGOs. Priority is given to producing numbers to justify expected results and short shrift is given to building local capacity. This is not a new observation, but it is a case of intractably perverse incentives (Holcombe).

(5) **There can be no naiveté that all African governments and civil societies are ready to exercise endogenous leadership.**

Not all African governments and civil societies are ready to exercise endogenous leadership that sets a development vision for the future, a vision that builds on traditions in order to root change in African soil. Corruption and patrimonialism flourish in many, but not all, places. The process can begin with those places where African traditions and knowledge have been used to build modern societies (chieftaincies in Botswana or Ghana) and where African professionals are now drawing on their own history.

(6) **Endogenous development is not about romanticising a dead past.**

Instead, it is about using the foundation of the strengths of the past and the present to ensure that Africans are in charge of the future they desire. A traditional African saying reflects the importance of grounding development in its context: "*There can be no tree without roots*." As with nearly every other country, Africa's past has many shadows and contradictions, but it also has many positive and developmental elements which have largely being marginalised and pushed to the periphery. There is organisational, developmental, and technical wisdom that can be adapted and adopted for modern times. When outsiders dismiss endogenous development approaches, they often do so by singling out the flaws in historical traditions (authoritarian kings) or current cultural practices (female genital cutting) and using these to characterise the whole of African cultural experience. In doing so, they are in a way perpetuating the colonial tradition of seeking to 'civilise' Africans. Several papers focus on looking for and using traditions that have promise as a foundation for new gender role definitions and practices (Msukwa, Galaa, Rwiyereka, Tinguery and Delaney.). Charles Banda's reflection piece suggests that development practitioners, both insiders and outsiders, need to hone listening skills in order to understand the essential meaning or purpose of traditional practices and how they might be adapted to meet modern challenges.

(7) **Endogenous development is not an either/or proposition but a "both" proposition**.

In the modern world, we need to build on the strengths of both the local and the imported but with the local (endogenous) taking lead. The African proverb tells us, "*the sympathiser cannot mourn more than the bereaved*". Blending modern knowledge and best practice with the African context and priorities needs to be done by Africans. Sympathisers can and should offer critical support, but the political, social, and cultural space for decision-making is African.

(8) **There is a general lack of a body of knowledge on best practice in endogenous development**.

Africans still seem to be stuck to our traditional oral culture, forgetting, as a traditional proverb tells us, that *"the palest ink is mightier than the strongest memory"*. Where a little body of knowledge does exist, there is a big challenge with access and distribution to a wider audience. One still observes that a large number of writers on the subject of endogeneity are non-Africans. Here, the African proverb, *"until the lions have their own writers, the story of the hunt will always praise the hunter"* is not so different from the Western proverb that the *"victor writes the history"*. There is a power differential between African development specialists and Western specialists in terms of access, funding, and research capacity. There is a great need to encourage a large number of African writers to get involved in documenting best practice in endogenous development; more importantly, to find ways to move this knowledge from the periphery to the centre of developmental discourse; and also to generally improve access and distribution mechanisms.

(9) **Development must always start from inside out and not outside in**.

This has implications for political, academic, and practice leadership. African leaders have the responsibility to lift endogenous development from community-level, discreet, small-scale practice to policy level. Policies and practice must be informed by African models of development, tested and rigorously evaluated. There is a lot of work that needs to be done to make such concepts as *ubuntu, botho, harambee, teren, ushadidi,* and *gacaca* robust enough to guide policy and practice. As of today, Africa still looks to the West for policy direction and best practice. This is also encouraged by the general reluctance of the West to accept and give space to African-originated developmental ideas. Due to geographical power differences and interests, it is relatively much more difficult to promote indigenous-originated ideas as compared to those originating in the West.

Looking forward

The lessons above imply their own solutions, and responsibilities for African leaders, practitioners, and scholars as well as for the donor community. Rather than focus on a list of changes that can be made in donor policies, development practices and research methods and priorities, we suggest that the move toward endogenously led and grounded development in African will be facilitated as much by the larger shifts taking place in the global approach to development as by the needed accumulation of changes in development policy and practice. The old world of development, led by development assistance and guided by an evolving set of policies (mostly economic), is faltering. On the negative side there is concern from disparate sources that these top-down policies have not been successful in reducing or sustaining the most severe poverty and inequality. At the same time, we do not yet have alternatives to the Western development ideologies of economic growth. Though some of the leadership in development policies has moved from traditional bilateral and multilateral agencies to foundations like the Bill & Melinda Gates Foundation, the development approaches remain top down and reliant on results-driven implementation that can lead to perverse consequences for learning and building local capacity (Fukuyama 2003; Ebrahim 2003; Saenbut 2014). On the positive side, recent decades have seen endogenously led development among large and medium-sized countries (such as China, Brazil) that have been successful in substantially reducing poverty and/or inequality. In Africa we see emerging successes in economic and social development in countries (such as

Rwanda, Malawi, and others) that have been driven by African leadership and sometimes by innovative uses of traditional practices to bolster effective policy implementation. We do not yet know whether these African successes are sustainable, and whether they can surmount continuing challenges of governance, conflict management, and effective development management. If success stories in Africa flourish, the ecology for endogenous leadership and approaches will also flourish.

References

Ebrahim, A. 2003. *NGOs and Organizational Change: Discourse, Reporting and Learning*. Cambridge: Cambridge University Press.

Fukuyama, F. 2003. "The Missing Dimensions of Stateness." In *State-Building: Governance and World Order in the 21st Century*, 1–42. Ithaca, NY: Cornell University Press.

Saenbut, N. 2014. "Participation and Ownership Rhetoric and the Reality of Development Practice: A Case Study." Unpublished master's paper, The Heller School for Social Policy and Management, Brandeis University, USA.

Index